ROUTLEDGE LIBRARY EDITIONS:
ENVIRONMENTAL POLICY

Volume 1

PROTECTING THE PERIPHERY

PROTECTING THE PERIPHERY
Environmental Policy in Peripheral Regions of
the European Union

Edited by
SUSAN BAKER, KAY MILTON AND
STEVEN YEARLEY

LONDON AND NEW YORK

First published in 1994 by Frank Cass & Co. Ltd

This edition first published in 2019
by Routledge
2 Park Square, Milton Park, Abingdon, Oxon OX14 4RN

and by Routledge
52 Vanderbilt Avenue, New York, NY 10017

Routledge is an imprint of the Taylor & Francis Group, an informa business

© 1994 Frank Cass & Co. Ltd

All rights reserved. No part of this book may be reprinted or reproduced or utilised in any form or by any electronic, mechanical, or other means, now known or hereafter invented, including photocopying and recording, or in any information storage or retrieval system, without permission in writing from the publishers.

Trademark notice: Product or corporate names may be trademarks or registered trademarks, and are used only for identification and explanation without intent to infringe.

British Library Cataloguing in Publication Data
A catalogue record for this book is available from the British Library

ISBN: 978-0-367-18894-8 (Set)
ISBN: 978-0-429-27423-7 (Set) (ebk)
ISBN: 978-0-367-18965-5 (Volume 1) (hbk)
ISBN: 978-0-367-18967-9 (Volume 1) (pbk)
ISBN: 978-0-429-19962-2 (Volume 1) (ebk)

Publisher's Note
The publisher has gone to great lengths to ensure the quality of this reprint but points out that some imperfections in the original copies may be apparent.

Disclaimer
The publisher has made every effort to trace copyright holders and would welcome correspondence from those they have been unable to trace.

PROTECTING THE PERIPHERY

Environmental Policy in
Peripheral Regions of
the European Union

edited by
SUSAN BAKER, KAY MILTON and
STEVEN YEARLY

FRANK CASS

First published in 1994 in Great Britain by
FRANK CASS & CO. LTD
Newbury House, 890-900 Eastern Avenue, Newbury Park, Ilford,
Essex IG2 7HH, England

and in the United States of America by
FRANK CASS
c/o International Specialized Book Services Inc.
5804 N.E. Hassalo Street
Portland, Oregon 97213-3644

Copyright © 1994 Frank Cass & Co. Ltd

British Library Cataloguing in Publication Data

Protecting the Periphery: Environmental
Policy in Peripheral Regions of the
European Union. – (Regional Politics &
Policy Series, ISSN 0959-2318)
 I. Baker, Susan II. Series
333.7094

ISBN 0 7146 4584 2 (cased) ISBN 0 7146 4114 6 (paper)

Library of Congress Cataloging-in-Publication Data

A catalogue record for this book is available from the
Library of Congress

This group of studies first appeared in a special issue on 'Protecting the Periphery: Environmental Policy in Peripheral Regions of the European Union in *Regional Politics & Policy* Vol. 4, No. 1, published by Frank Cass & Co. Ltd.

All rights reserved. No part of this publication may be reproduced in any form or by any means, electronic, mechanical, photocopying, recording or otherwise, without the prior permission of Frank Cass and Company Limited.

Typeset by Vitaset, Paddock Wood, Kent
Printed in Great Britain by
Watkiss Studios Limited, Biggleswade

Contents

Environmental Policy and Peripheral Regions of the European Union: An Introduction	*Steven Yearley, Susan Baker and Kay Milton*	1
Policy Co-ordination in Brussels: Environmental and Regional Policy	*Sonia Mazey and Jeremy Richardson*	22
Policy Networks on the Periphery: EU Environmental Policy and Scotland	*Elizabeth Bomberg*	45
Administrative Capacity and the Implementation of EU Environmental Policy in Ireland	*Carmel Coyle*	62
National Environmental Policy-making in the European Framework: Spain, Greece and Italy in Comparison	*Geoffrey Pridham*	80
Spanish Pollution Control Policy and the Challenge of the European Union	*Susana Aguilar-Fernández*	102
Environment and the State in the EU Periphery: The Case of Greece	*Maria Kousis*	118
The European Union and Visegrád Countries: The Case of Energy and Environmental Policies in Hungary	*Janne Haaland Matláry*	136
Ups and Downs of Czech Environmental Awareness and Policy: Identifying Trends and Influences	*Petr Jehlicka and Jan Kara*	153

Environmental Policy and Peripheral Regions of the European Union: An Introduction

STEVEN YEARLEY, SUSAN BAKER and
KAY MILTON[1]

INTRODUCTION

The course of the development of the European Union (EU)[2] through the 1990s appears increasingly uncertain, but one oasis of comparative certainty is that the institutions of the EU will play a growing role in influencing environmental quality in the member-states and in countries which aspire to membership. There is a variety of reasons for expecting this to occur. For one thing, many environmental problems plainly exceed the geographical and legislative limits of particular nations; the logic of supranational action is actively promoted by EU institutions and is increasingly acknowledged throughout the Union. Moreover, environmental improvement measures (notably pollution control and the protection of the natural heritage) often command public support and approval and this gives the agencies of the EU an additional interest in aligning themselves with such measures. Thirdly, the Union has recently committed itself to the pursuit of sustainable development, thus explicitly integrating environmental considerations into its core mission – the promotion of socio-economic development. While ostensibly a new policy development, this commitment to sustainability is, in an important sense, the acknowledgement of a relationship which has existed all along; namely, that economic development and environmental policies cannot properly be treated in isolation, because the most significant threats to the European environment have come about precisely through economic development often financed by the EU, whether through the pursuit of greater agricultural productivity, developments in transport infrastructure, loss of habitat to commercial developments or increases in power generation. In other words, the EU has so far played its largest role in shaping the environmental condition of member-states, not through its environmental policy measures, but through the consequences of its free-trade, economic development and agricultural enhancement provisions. A policy of fostering sustainability has to begin by acknowledging this fact.

The papers in this collection focus on one significant but relatively neglected aspect of the EU's impact on the environment. They examine the influence of the EU on environmental quality and on environmental policy in the peripheral areas of the Union. The studies which follow indicate that this influence is of practical and intellectual interest for several reasons. For one thing, virtually by definition, the periphery is more in want of economic development than the core. Accordingly the trade-off between development and environmental protection, or (to put it another way) the degree of difficulty of introducing sustainable development, is likely to be most intense in peripheral areas. Second, often because of their peripheral position, the areas of the EU's periphery contain large regions of characteristic and relatively undamaged habitat which render them of ecological significance to the Union as a whole. Third, the EU-wide environmental standards adopted may not seem appropriate to development agencies in the periphery, or the standards may not be seen as legitimate even if they are accepted as the 'rules of the game'. For this reason, EU environmental policy may be viewed ambivalently in peripheral regions and compliance and implementation may be patchy. This may also occur, indeed it often does occur (as the following chapters indicate), because peripheral regions suffer from institutional fragmentation or lack the administrative capacity necessary to introduce and enforce environmental standards. Lastly, emerging environmental interest groups in the periphery may look to the EU as a particular ally in advancing their campaign goals; changes in EU provisions may thus have a large impact on the strategy and fortunes of campaigners at the periphery.

The chapters which follow have been chosen to throw light on these issues in a variety of ways. The selection includes analyses of the policy process at the level of EU institutions (featured prominently in the papers by Bomberg and by Mazey and Richardson) as well as containing a series of case studies of peripheral areas within the EU. There are also two studies of the influence exerted by the EU on environmental policy in countries on the external periphery of the Union; the EU is in a position to influence such countries because they aspire to EU membership or wish to trade extensively with the EU and therefore feel compelled to meet EU product standards, or because the EU is actively involved in shaping their environmental policies through its PHARE Programme (Poland and Hungary Action for Restructuring the Economy)

The role of this chapter is to provide an introduction to EU environmental policy and to highlight significant themes which recur throughout the case studies which follow.

DEVELOPMENTS IN EU ENVIRONMENTAL POLICY

As already mentioned, environmental policy is a significant and rapidly-growing concern of the European Union. While the original Treaty of

Rome (1957) contained no reference to the environment or environmental policy, the European Community (EC) has acted in this policy area more or less from the outset and with greater vigour since 1972 [for an authoritative overview see *Haigh, 1992: 233–49*]. Such action was initially justified primarily in relation to the elimination of discrepancies which affected countries' competitiveness; for example, if a country maintained lower standards of environmental testing than other member-states, that could give it an unfair economic advantage. However, in the absence of explicit agreements on positive policy towards the environment, there was always a danger that policy interventions in this area would lack consistency and be more than usually subject to political whims.

Some order was introduced into this rather *ad hoc* and unsystematic arrangement after an EC Heads of State meeting in Paris in 1972; from 1973, the Commission began producing four-yearly Environmental Action Programmes which identified the developing principles governing environmental policy and which established priorities for environmental action. For instance, the Third Programme, which ran from 1982–86, emphasized preventative approaches to pollution control and saw the introduction of Environmental Impact Assessments as a legislative measure intended to institutionalize a degree of preventativeness. The fifth such Programme was published in March 1992. It formally introduced the principle of sustainable development and directed action towards the integration of environmental concerns into all areas of community policy-making. For illustrative purposes we can note that this Programme's objectives included the promotion of energy efficiency, improvements in environmental information for consumers and the establishment of an 'environmental policy implementation group made up of both Commission and member state government officials' [*Croner's Europe, July 1992: 2-553*].

EU environmental policy thus developed more or less as a by-product of economic and social policies. Indeed, environmental policy was so lacking from the list of central priorities that it was only finally granted a formal status within the Treaty through the Single European Act (1987) which consolidated recent expansions in the objectives of, and principles underlying, policy. The preservation and protection of the environment were adopted as official Community objectives alongside the existing commitment to environmental measures aimed at enhancing human health.

The standing of EU environmental policy has also been decisively influenced by the way in which policy-making has been undertaken. While occasional use has been made of regulations and decisions in introducing environmental policies (for example, in the case of international conventions such as the Barcelona Convention, protecting the Mediterranean Sea, and CITES, the Convention on International Trade in Endangered Species of Wild Flora and Fauna), most provisions have been legislated for using directives. Unlike regulations and decisions, which have a directly binding force, directives have to be interpreted by the nation in which they are to be implemented; as Budd and Jones express it, 'though binding on member states as regards the results to be achieved, the form and method of achieving these results is left to the discretion of national authorities' [*1989: 36*]. Directives are also, as far as the EU itself is concerned, a low-cost route towards achieving environmental regulation. Thus, although rapidly growing in importance, environmental policy-making in the EU has developed in such a way that it allows for an unusually high degree of discretion on the part of member-states.

Indeed, two steps are required to put directives into force at the member-state level. The first is that each member-state must pass the appropriate national legislation, a legal process referred to as 'enactment'. The second is that the national legislation must be applied and enforced in practice, a complex process usually described in policy studies literature as 'implementation'. In the field of environmental policy, enactment has been considerably delayed by many member-states and implementation, once enactment has been achieved, remains highly unsatisfactory, across a wide range of member-states as well as policy areas [*Siedentopf and Ziller, 1988*].

While some commentators have viewed EU environmental policy as something of a success, at least by comparison with other areas of policy-making [*Lasok and Bridge, 1989: 469*], it is sobering to note that even enactment has not always proceeded smoothly. Thus, both Italy and Britain have been taken to court over failures to enact directives and, as we shall see, the very flexibility of directives as a policy tool has allowed some countries to escape complying in spirit, even if they have followed the letter of the law and actually enacted European legislation. The adoption of this regulatory style for environmental matters makes the issue of compliance truly central.

Such displays of national reluctance to participate in environmental reform have been offset to some extent by EU institutions' own propensity for environmental activism. This largely stems from the fact that policy interventions to promote environmental benefits can readily be presented as in the public interest and have tended to be popular. This has been one route by which the EU can increase its legitimacy. Moreover, the formation in 1981 of DGXI, the Directorate General with primary responsibility for environmental legislation, ensured that one component of the EU administrative machinery has an institutional interest in environmental reform. Of course, other Directorates General, such as those for Transport and Agriculture, were also highly influential in shaping the European environment. But DGXI, often seen as heavily influenced by the environmental lobby [*Mazey and Richardson, 1992a*], was concerned to ratchet environmental standards upwards, an issue explored further in Mazey and Richardson's chapter. Furthermore, EU officials have been keen to build relationships with regional authorities as well as national agencies, thus providing themselves with ways of circumventing recalcitrant governments.

EU environmental activism has also been spurred because environmental policies commonly affect issues which themselves have a transnational character. For example, while EU institutions can readily justify an interest in working hours or in educational qualifications, since these are likely to affect the prices of goods or the mobility of labour, the policies themselves have an impact primarily within individual member-countries. By contrast, many – though by no means all – environmental issues have direct effects outside the country where they originate. Whether the issue is acid emissions, pesticides entering waterways or contributions to the enhanced greenhouse effect, one country's policy will directly influence environmental conditions in its neighbours. Recognition of the international character of the environmental crisis, especially since the 1980s, has resulted in a questioning of the conventional wisdom that the nation state is the most appropriate level for dealing with the environment [*Weale, 1992: 28*].

To argue in this way is not to overlook other interests which may favour the adoption of environmental policy standards. The shift away from the level of the nation-state towards a centralization of decision-making within the EU also benefits much large-scale European industry [*Majone, 1991. 26*]. Social analysts have often assumed that industry's natural orientation is to oppose environmental regulation. Within the EU at least, the prospect of common standards has proven to be more attractive to the majority of business than somewhat lower average standards which vary from one country to another and may be subject to unpredictable changes. Of course, this appeal is not universal. Typically, the larger the firm, the more it stands to benefit from a single unified market. Furthermore, firms in countries with a propensity for higher environmental standards (such as the Netherlands or Germany) may find rising general standards of particular benefit since they are already operating at the level of likely new regulations. By and large, peripheral areas are likely to be adversely affected by this process, as Pridham shows in his chapter.

Finally, it is important to emphasize that EU environmental policy, despite political and commercial pressures favouring convergence, is not a unified entity. For example, as Mazey and Richardson point out in their chapter, different Directorates General have been responsible for environmental policy and for regional development; their overlapping agendas have not always been in conformity. Thus, the problems of devising and implementing an environmental policy do not derive only from the complexity of environmental policy formulation but stem also from the organizational structure of the EU itself.

POLICY-MAKING AND IMPLEMENTATION IN THE EU CONTEXT

The interest in 'policy implementation' arose in the United States in the 1970s as a result of what were perceived to be the failures of policies to deal with the major problems of that period, including racial discrimination and poverty. Initially, the approach adopted was what became known as the 'top-down' model of policy implementation. This model saw policy-making as a rational process in which policy was an almost readily identifiable entity. To fill the implementation deficit (policy failure), according to this model, meant improving the mechanisms by which it was transmitted by bureaucrats: clearly and unambiguously defining the 'policy', removing interference, shortening the chain of transmission and so on [Ham and Hill, 1985]. In the 'top-down' approach, it was thought that the different stages of the policy process (formulation, implementation, evaluation, termination) could be easily distinguished.

This top-down understanding of the policy process was challenged by what became known as the 'bottom-up' approach. Central to this new approach was the belief that the top-down model presented an unrealistic understanding of the complexities of the policy-making process. Rather than being a rational, technocratic process, policy is seen, in the bottom-up approach, as the outcome of a pluralistic and political bargaining process. Bargaining takes place within a wider political context which incorporates an array of different, and often competing, interests and actors. This approach considers that policy, far from being readily identifiable, is a slippery concept; policy-makers are perceived less as machines and more as complex networks and groups bargaining towards policy resolution.

In this perspective, far from being a mere mechanical undertaking at the end of a series of discrete steps, implementation is seen to be closely connected to policy formulation. The implementation of policy loops back in such a way as to reformulate the policy itself. Furthermore, research on 'street-level' bureaucrats – the police officer on the beat, the social security clerk behind the counter – has revealed that 'policy' is what happens at this level rather than what is defined by those at the top [*Lipsky, 1980*]. The difference between the two has been shown to be very considerable.

In the 'bottom-up' approach, the political dimension is recognized as an important variable in policy-making. Politics here means the pluralistic bargaining processes characteristic of democracy as well as party politics. Thus, implementation involves inter-organizational relationships, with networks of multiple actors who operate to achieve policy objectives or goals. According to Pressman and Wildavsky this process of interaction forms a seamless web [*Pressman and Wildavsky, 1979*]. This approach has led to the study of policy networks and policy communities, and the use of network theory to try to understand the sets of interests involved in both the formulation and the implementation of policy (see the chapters by Bomberg and by Mazey and Richardson).

This shift in theoretical approaches has implications for our understanding of the implementation deficits referred to above, a main focus of attention in this volume. Initially, in the top-down approach, a deficit arose whenever a gap opened up between the initial policy intention and the actual policy output. However, following the insights of the bottom-up approach, the concept of implementation deficit takes on a more diffuse and subtle meaning. Policy failure can no longer be seen as a break in this chain of command nor rectification as involving a fine tuning of the implementation system.

In the EU context the bottom-up approach is very relevant, especially the focus of attention on policy communities and policy networks. In particular, it provides a model that can accommodate the complexity of the EU policy-making process, which operates at both the formal and informal levels. The formal level is, in any case, well known for being highly fragmented and complex. Proposals are drafted by the Commission and these may have inputs from the advisory committees and be subject to political scrutiny by the Commissioners' cabinets. The Commission's proposals then go forward to the representative bodies: the Parliament (EP), the Economic and Social Committee (ESC) and, under provisions in the Maastricht Treaty, also to the Regional Committee (RC). The responses of the representative bodies are then examined by the Council. Here attention is given to both the original proposal and the responses to it. Council may then have to negotiate over the proposal; so too, as a last resort, may the Heads of State. Directly binding decisions are implemented by the Commission and the indirectly binding ones (directives) go to national governments for implementation. Besides these formal links, there are also informal links with member-states, as each member of the EU is entitled to a certain proportion of Commissioners, statutory civil servants, representatives, EP seats, ESC seats and RC seats, and a seat on the Council as well as the Court [*Van Schendelen, 1993*]. Thus the process involves intense interaction and feed back with member-states. As Bomberg has indicated in her chapter on Scotland, regional interests may or may not be included in this interaction.

Moreover, the above outline gives us the linkages only in the formal policy-making process. In practice, according to Van Schendelen, this machinery functions in an even more bottom-up manner. He argues that the EU civil servants can act, despite the European Statute they have signed, with a double or even a single loyalty in favour of specific nations, sectors, ideologies or regions. Furthermore, national publics or private interest groups can succeed in 'parachuting' their people into the Commission or placing them in useful nearby locations. Membership of policy-making bodies, as well as loyalties, may overlap. National interests and characteristics can easily be imported into the machinery and thus influence decision-making. The EU policy-making process is thus the target of intense lobbying [*Van Schendelen, 1993*].

Mazey and Richardson have argued that there are sound reasons why environmental groups form and join lobby groups, known as Euro-groups, and that these relate directly to the nature of the EU policy process [*Mazey and Richardson, 1992b: 1–5*]. Policy-making within the EU is dispersed across a number of institutions and there are several policy initiators. Their research into that policy-making process has found that it is best described as 'loose' and 'open', where participation by groups is unpredictable and where policy ideas may appear suddenly and from little-known sources. They have argued that the agenda-setting stage of the policy process of the EU is unstable and exhibits a high degree of unpredictability. To add to these difficulties, 'policy-making is highly compartmentalized with little horizontal co-ordination between different Directorates General which have a shared interest in an issue' [*Mazey and Richardson, 1992b: 7*]. As a consequence they argue that 'keeping track of EU policy initiatives is a major undertaking for groups many of whom lack sufficient resources to perform this task on their own' [*Mazey and Richardson, 1992b: 3*]. This acts as a major reason why groups, be they environmentalists or industrialists, form and join Euro-groups.

The Commission tends to be the focus of most EU lobbying. For their part, the formation of Euro-groups brings distinct advantages to the Commission and this further encourages their growth. To begin with, Euro-groups can provide the Commission with much needed specialized knowledge. Second, the Commission's negotiating hand can be considerably strengthened with the Council if it can show that its proposals are supported by organized, often influential, interests. Furthermore, by presenting the Commission with broadly united and coherent policies, Euro-groups can save the Commission from becoming embroiled in ideological, sectoral or national interests [*Nugent, 1991: 233–34*]. Finally, the Commission is small and has limited resources and as a consequence it is dependent upon outside sources for expertise and information and 'on the ground' information. In this context, argue Mazey and Richardson, it is in the Commission's own interest to deal with Euro-groups, especially if they are at the same time both *representative* and *expert*.

Some care needs to be taken with such a pluralist approach lest it implies an equality of influence which is at odds with the facts. On this view of policy-making there exist various stages during which the peripheral regions or member-states can inject their concerns. However, in the open system of competition, where national and sub-national interests, private lobby groups and firms try to influence the content of EU policy, peripheral regions are highly disadvantaged, given their weak resource base. The lack of a regional dimension has been pointed out, for example, in the criticism made of the Structural Funds, which have also been severely criticized for their lack of adequate environmental safeguards [*Baker: 1993a*]. Peripheral member-states are similarly at a disadvantage. Thus in the EU policy making process, some member-states have greater weight, not just in the technical sense of voting, but also in the more meaningful sense of political and economic 'clout'. Some member-states can therefore be seen to be leaders in the policy field, such as Germany, undoubtedly setting the agenda of European Monetary Union (EMU); Germany is also a key influence behind the adoption of a strong regulatory approach towards environmental policy within the EU.

Furthermore, the weak resource base of peripheral regions and member-states limits their capacity to implement policy. Even if policy were clearly defined, as the top-down model claims, by the time it reached the peripheral regions, its chances of being implemented according to the original intentions would be very slim. This is especially so in the field of environmental protection, requiring as it does high levels of scientific and technical expertise, sometimes of a novel character [*see Yearley, 1992*].

The complexities of the EU policy-making process are all the more evident in the field of environmental policy. Despite the declarations of the Fifth Action Programme and the Maastricht Treaty, environmental policy remains in competition with many of the central policies of the EU, especially agriculture and transport policy. As a consequence, there is fragmentation and contradiction at the very heart of the EU policy system [*Baker, 1993b*]. There is as yet no peripheral regional dimension to environmental policy, although under the Maastricht Treaty this may change. Putting this into practice may well prove difficult, as there is a tension with Union policy between harmonization and differentiation at the regional level. The principle of subsidiarity, also mentioned in the Maastricht Treaty, is proving to be an even more contentious issue.

ENVIRONMENTAL POLICY ANALYSIS AND THE CONCEPT OF 'PERIPHERY'

If there are sound reasons for examining the dynamics of EU environmental policy, are there also good grounds for thinking that environmental policy at the periphery will be worthy of interest? Our argument is that there are. For example, many of the most valuable and undisturbed habitats in the EU are to be found at the periphery, where they have often continued to exist precisely because of traditional agricultural practices or because of the absence of industrial development and a modern transport infrastructure [*Baker, 1993a*]. In effect, the environmental value of these locations has been maintained by default. Still, the conservation of such areas may therefore be at odds with their economic development; the resulting potential for conflict gives environmental policy in the periphery a definite piquancy. EU commitments to socio-economic development and to environmental protection threaten to collide [*Baker, 1993b*]. Furthermore, 'peripherality' may be demonstrated in administrative terms as well as in economic, geographical and cultural ones. As pointed out earlier, and as shown in, among others, the chapters by Coyle, Pridham and Aguilar-Fernández, peripheral areas may lack the organizational and administrative capabilities which are assumed to be in place at the centre. In turn this may lead to unanticipated problems with implementation and the delivery of policy objectives.

Although this issue appears tangible and significant, there is a conspicuous risk of circularity for any analyst who tries to make sense of it. It is tempting to explain the acute environmental problems of peripheral areas in terms of their peripheral status *at the same time* as one uses environmental characteristics as a means of identifying which areas are to be classed as peripheral. This clearly will not do. We need some more or less independent method of recognizing peripherality, so that we can check whether or not peripheral areas do have distinctive ecological problems and environmental policy needs.

The search for this method of identifying the periphery is far from easy. 'World-systems' authors who divide the world into core, periphery and semi-periphery [*Martin, 1990; Wallerstein, 1979*] tend not to agree about the exact grounds for allocating countries to one category or another (see Kousis' chapter for a discussion of these labels). Worse still, both countries and smaller units within (or across) countries can be referred to as peripheries. Portugal has fairly decisive claims to geographical peripherality within the EU; Greece possibly even more so since its nearest neighbours are all non-EU countries. Yet within both these nations there are greater and lesser peripheral areas. The more traditional southern parts of Portugal have tended to be at odds with the north and with urban areas and can plausibly be seen as Portugal's periphery. Even more credibly, the groups of Greek islands can lay claims to being on the geographical periphery of the European periphery. The picture is further complicated since even within the 'core' there are peripheries. Thus, the UK has very pronounced peripheral regions in northern Scotland and Northern Ireland. France has its 'Celtic' fringe too, and geographically and politically remote Corsica; and, in a very different sense, Germany has a poorly-developed periphery in the former German Democratic Republic.

Furthermore, one can usefully speak of a periphery *beyond* the EU, embracing many nations more or less keen to join. Some of these, such as Switzerland and Austria, are ironically closer in geographical and cultural terms to the EU's administrative core than some of its member-states. Others, such as Poland, are peripheral in an economic as well as a geographical sense. This 'periphery' extends far into the former USSR and into eastern European countries such as Hungary and the Czech Republic (considered in the chapters by Matlary and by Jehlicka and Kara). Countries which aspire to membership or which need to trade extensively with the EU feel the effects of their peripherality very deeply. For example, they may well attune their laws and trade practices to conform to EU standards for fear of disqualifying themselves from membership, even if these standards are not demanded domestically. Through the PHARE Programme they may even have their environmental and economic policies directly affected by EU initiatives, initiatives which, at least in part, are aimed at reducing the impact of eastern pollution on EU member-states. Lastly, there are countries on the southern edge of the EU which are also influenced by their proximity even if they are not immediately planning an alliance. Their environmental regulations may be affected by trading policies, by standards demanded from their EU trading partners, and through programmes designed to influence the environmental quality of the Mediterranean.

On the face of it, an alternative approach would be to adopt designations used by administrative authorities. Thus, the EU itself designates five Priority regions, including Objective One Priority Regions, and there are formal definitions of less favoured areas and of 'Cohesion Countries' which qualify for the EU's recently introduced Cohesion Funds. Within member-states also there are designations relating to regions with special socio-economic needs. The disadvantage of adopting these official characterizations is that they tend to be drawn on a rather gross scale, treating, for example, virtually the whole of Northern Ireland as a border region, including areas some 100 kilometres from the border with the Republic of Ireland. Moreover, since there are economic and political advantages as well as disadvantages to being designated in these ways, actual designations may owe as much to political manoeuvring as to objective assessment. Former UK Prime Minister Thatcher was famously reluctant to have areas of Great Britain designated as Objective One, whatever attractions this may have held for regional authorities. Official categories cannot simply be taken over and adopted for analytic purposes.

The approach adopted in this volume is essentially iterative. The case study countries were selected on both socio-economic and geographical grounds. Authors met and discussed their analyses at a Workshop organized by the European Consortium for Political Research (in Leiden, April 1993). The editors and authors then used the comparative case study material to refine their interpretations of peripherality for the final versions of the chapters. As elsewhere in the social sciences, such iterative methods mean that there is always room for further refinement.

CULTURAL DIVERSITY AND ENVIRONMENTAL POLICY

Fundamental to the understanding of policy processes in any context is a knowledge of the political culture within which policies are both formulated and implemented. Given that the EU includes a range of countries with diverse cultural traditions, it is to be expected that cultural variations will influence any imbalances, in the operation and outcomes of policy processes, that might be observed both among member-states and at the regional level within any one state.

Is it possible to identify cultural differences which correspond to the distinction between centre and periphery, and which might, therefore, help to account for the nature of policy processes at the periphery? This degree of generalization is more than many social scientists would care to attempt. However, one branch of cultural theory, deriving from the work of Douglas [*1970; 1987*], offers some tentative insights.

Douglas presents a model of cultural diversity which is based on the premise that different institutionalized ways of thinking and acting are generated by different forms of social organization [*Douglas, 1970: 86–91*]. Where social organization is dominated by corporate groups, cultural perspectives will tend to vary along a continuum between the hierarchical and the egalitarian, depending on the degree of constraint or independence experienced by individuals in their own actions. Where social organization is dominated by ego-centred networks, cultural perspectives will tend to vary between the entrepreneurial and the fatalistic, again depending on the degree of independence experienced by individuals. This model has been used as a framework for analysis at different levels; in the classification of whole societies [*Douglas, 1970*], of different sectors within a culturally diverse society [*James* et al., *1987*], of different levels within a single organization [*Thompson and Wildavsky, 1986*], and of different policies formulated by a single government [*Milton, 1991*].

We would not wish to suggest that this model can fully *explain* cultural differences between the core and the periphery of the EU, or of any other culturally heterogeneous organization. Nevertheless, it might be used to suggest fruitful lines of comparison. For instance, environmental policies emanating from the EU appear, on the face of it, to be guided by egalitarian and hierarchical principles; they are aimed at equalizing environmental standards throughout the Union and operate through mechanisms which require a degree of central control. This is precisely what we might expect in terms of Douglas' model, given that the core states within the EU, those which exercise the most influence over Union policy, are characterized by corporate forms of social organization, both in their commercial sectors and in their environmental lobbies. Some of the peripheral states, on the other hand, particularly Greece, Ireland and Italy, are presented in the literature as characterized by a network structure, with political systems traditionally dominated by patronage and clientelism [*see Campbell, 1964; Komito, 1989; Lewanski, 1993*]. This raises the question of whether some of the problems involved in putting EU environmental policy into operation at the periphery, problems which generate an implementation deficit, can be understood in relation to a fundamental incompatibility in political culture. This kind of analysis is implied in some of the arguments presented by Coyle, Pridham, Aguilar-Fernández and Kousis, below.

A cultural incompatibility of a different kind is addressed in the chapters by Matlary and by Jehlicka and Kara. The communist regimes of eastern Europe are known to have had severely damaging impacts on the environment, dominated, as they were, by a political ideology which cast humankind as the conqueror of nature and which consistently valued the pursuit of economic goals more highly than environmental protection. The cultural legacy left by the communist systems makes it difficult for countries like Hungary and the Czech Republic to support and implement policies which appear to give priority to environmental protection, or which seem to suggest that nature is, in its own right, worthy of consideration. In the next three sections we use our understanding of peripherality in the EU to discuss leading features of the dynamics of environmental policy at the periphery.

PRESSURES FOR ENVIRONMENTAL REGULATION AT THE PERIPHERY

The studies in this volume indicate that peripheral areas are feeling a pressure to raise environmental standards to comply with EU targets and legislation. The principal spur to environmental reform is a desire to conform to EU standards. This assertion risks sounding like a truism, but it is not. The point is that the prime motivation to raise standards is not to improve environmental quality *per se* but to be seen as good Europeans (as Pridham suggests in his chapter) or to avoid censure, to qualify for advantageous funds or, in the case of Hungary and the Czech Republic in particular, to be seen as suitable trade and political partners.

Of course, as we shall describe below, there are pressure groups and local interests in the peripheries which do wish to raise environmental standards as such. In some cases, especially those involving habitat loss or concentrated air pollution in certain urban settings, environmental degradation may be terribly apparent and it may be in the economic or health-related interests of groups to press for environmental improvement. But just as often, the motive for environmental reform is a concern not to fall too foul of the EU, or a desire to strike a deal with fellow member-states. In so far as improved environmental standards are not pursued for their own sake but arise from horse-trading and political deals, countries may have an interest in passing suitable-looking laws but then pursuing implementation only unenthusiastically. Some evidence that this may be the case was provided in *The Economist* (20 July 1991: p.54) where it was shown that Spain, Italy and Greece had much the highest numbers of outstanding breaches of EC environmental rules (over 45 per cent of the breaches for all 12 member-states). Some details of the interplay between politics and policy response are examined in the next section while the case study chapters (particularly those by Kousis, Coyle and Aguilar-Fernández on Greece, the Irish Republic and Spain) examine the different constellations of reasons in different cases.

THE POLITICS OF PERIPHERIES' ENVIRONMENTAL POLICY

As mentioned at the start of this chapter, a prime consideration in peripheral areas is that environmental protection is often viewed as having a particularly sharp trade-off against economic development and employment creation. Many policy-makers in the core areas would now claim that the apparent opposition between environmental protection and economic advance can be transcended since, for example, prosperity often flows to those firms which make sparing use of natural resources and which are known to respect environmental quality (the so-called 'ecologically modernized' firms). This happy state has not been attained in the periphery where arguments about the economic and employment 'costs' of environmental measures can still be regularly heard (see, for example, Pridham's discussion of the Spanish response to the proposed carbon tax). Furthermore, groups in the peripheral areas may view EU initiatives aimed at conserving traditional landscapes or preventing the pollution of sensitive areas as measures which benefit the core (by, for instance, securing attractive recreational sites), at the same time as impeding the development of the periphery. Of course, the periphery is not always cast as the victim; Pridham shows how Majorca has pioneered policies on sustainable tourism. Still, the case studies suggest that many environmental protection measures may be construed as instances, not just of the core failing to help, but actually of the core exploiting the periphery.

This problem which, however old-fashioned the arguments may seem, features regularly in the case study materials, is compounded by an important administrative characteristic of peripherality. The implementation of environmental policy can be particularly problematic in peripheral areas since environmental protection requires complex administrative and managerial developments, co-ordination between numerous state and quasi-state agencies and the involvement of a large and complex set of actors. This is as true for agricultural management policies as it is for planning or industrialization. Such co-ordination may be hampered because of inadequate administrative structures, institutional fragmentation, lack of resources and staff-power to commit to environmental management, or because actors do not treat environmental concerns as matters of priority (see, in particular, the chapters by Coyle and Aguilar-Fernández).

The introduction of increasingly far-reaching environmental policies will impact on established relationships between peripheral states' governments and their regional authorities. They will also affect the relative positions of public authorities and the private sector (by increasing regulation and inspection for example) and may contribute to changing relationships between employers and unions. Given the flexibility which is associated with directives, these various actors are able to mobilize their resources to influence the details of the way the directives are implemented. In some peripheral areas, notably the Irish Republic, the domestic private sector has been relatively weak while foreign investment has been able to command a good deal of strength. Changes in environmental policy may well strengthen governmental bodies and thus work contrary to the perceived interests of foreign capital. In all, care must be taken to acknowledge the fact that the EU policy process is both complex and fragmented. The Commission has the monopoly on the proposal, the Council on disposal and the member-states on the implementation of policy. Overall responsibility for implementation rests with different types of political and administrative structures across the Union, with unitary, regionalized and federal states allocating both tasks and responsibilities differently. In contrast, actual implementation usually takes place at the local level, irrespective of the type of responsibility arrangements that exist.

A further major factor influencing the politics of policy at the periphery is that the treatment of environmental regulations can become entangled with other political and administrative concerns. This possibility is far from unique to peripheral areas as has been demonstrated by the interaction between the UK's environmental and privatization policies. This interaction has influenced the UK's approach both to the abatement of acid emissions and to issues of water quality. In the former case, for example, private electricity companies were allowed to switch to gas fuels (with cheaper abatement costs) even though the UK had used an assumption that the (formerly) state-owned electricity company would have to continue to burn coal as a bargaining counter in pressing for soft acid-emission reduction targets. But these complex interactions are even more likely in peripheral areas. The reasons for this are twofold.

First, it is because many areas of policy are undergoing change at the same time as the peripheries are feeling the effects of EU-inspired infrastructure development and the Single European Act. In particular, Cohesion Funds and targeted development funds – despite the addition of environmental considerations into the funding criteria – are adding a new dimension to the tensions that the periphery is experiencing between the imperative of increased economic development on the one hand and the new concern for environmental protection on the other [see Baker, 1993a].

For example, poorer member-states are receiving funds for road building and infrastructure development at the same time as they are being enjoined to conserve valuable habitat. Or, to take a recent example (*The Guardian*, 25 January 1993: p.20), the Spanish authorities may use European funds to build ecologically harmful dams to solve their water 'problem', a problem which only exists in relation to the aim of continuing to expand agricultural production (thereby adding further to food surpluses). The two policies may be contradictory in practice. This point is analytically straightforward but of clear practical significance, as some of the following chapters, including those by Matlary and Pridham, indicate.

Secondly, it is because environmental issues may become embroiled in some of the political disputes characteristic of peripheral areas, disputes which often bear on cultural and political identity. For example, groups in peripheral areas may place particular emphasis on managing their own landscapes and habitats as part of an attempt to assert control over their cultural heritage. Just when EU legislation is tending towards the introduction of uniform standards of environmental management, they may wish to manage local resources in an autonomous and unregimented manner (though it should be noted that EU institutions are not unsympathetic to this ambition).

Such complications are likely to be further aggravated because of current changes in the relationship between national and regional authorities within peripheral areas. Both sets of authorities are using the changes associated with the deepening of the EU integration process to try to reshape their powers and responsibilities. Terms such as 'subsidiarity' and 'partnership' are being employed to open this issue up for discussion. Environmental policies form just one component of these overall negotiations and power struggles, so that people's and politicians' responses to EU environmental policies and initiatives are affected not only by the perceived value of these measures themselves but also by the role these policies play in the overall power struggle.

This tension is particularly noticeable in the area of environmental policy. On the one hand, in response to the increase in both the amount and the complexity of EU environmental policy, many member-states are centralizing responsibility for environmental policy by establishing new, national-level bodies, for example, the Environmental Protection Agency in Ireland (see Coyle's chapter). This, as the case in Ireland reveals, has been at a cost to local authorities, who have lost a number of functions and responsibilities in an area of growing public visibility and importance. On the other hand, as mentioned earlier, regional forces within the EU, encouraged to a large measure by the Commission, are trying to by-pass the nation-state level and deal directly with EU institutions. The Maastricht Treaty appears to reinforce this tendency, though the significance of the so-called 'Committee of the Regions' is still being negotiated.

SOCIO-POLITICAL PROCESSES FAVOURING ENVIRONMENTAL REGULATION

To accept the foregoing points is not to suggest that the periphery's experience of EU environmental policy is exclusively that of the imposition of restrictions. If one takes the Large Combustion Plant Directive, for example, which guides the reduction of SO_2 and NO_x emissions up to the year 2003 (with overall target reductions of 57 per cent and 30 per cent respectively), we find that it actually allows for conspicuous growth in pollution at the European periphery. Greece and the Irish Republic are allowed to continue with enhanced SO_2 emissions compared with the 1980 baseline and are not required to cut 1993 levels at all during the subsequent decade. Portugal's NO_x pollution is even allowed to grow during the planned reduction period [*Skea, 1990: 19*].

Moreover, environmental campaigners and community groups operating in peripheral areas have often welcomed EU legislation since it provides leverage and ammunition which they have been unable to generate domestically. Such groups often look to EU standards for a tightening of domestic environmental law and they invoke the experience of other countries to show that, for example, alternative energy generation can be made viable or that air pollution monitoring can be effective. Indeed an argument can be plausibly made that, increasingly, environmental groups within member-states are acting as watch dogs of the EU policy process [*Mazey and Richardson, 1992b*].

Though environmental NGOs have tended to be initially weak in peripheral areas, we argue that deepening European integration has strengthened them in two ways: first by making these areas of greater interest to established campaign groups (for example, Greenpeace has an interest in pursuing 'dirty' industry throughout the EU and will subsidize operations in peripheral areas out of central funds) and, second, by giving them new legal instruments to apply. NGOs will also tend to be assisted by the internationalization of the movement, with groups being established explicitly at a European level: for example, the European Environmental Bureau and the Brussels offices of Greenpeace and European Friends of the Earth. However, these same measures threaten to lose campaign groups local support as they are seen to identify less with their traditional domestic constituencies and rather more with the international environmental movement. The adaptive response of peripheral NGOs is examined in several of the chapters but notably by Mazey and Richardson and by Jehlicka and Kara.

Growing awareness of a European identity has, in any case, allowed campaigners to try to shame polluters and other environmental miscreants by comparison with accepted standards of conduct in core countries. But rhetorical comparisons come a poor second to legal directives as a campaign tool. Campaign groups are more likely to meet with success in pressing a reluctant government to enact and implement EU legislation than in pressuring a government to meet standards which it should merely be ashamed of violating.

CONCLUSION: THE PERIPHERIES, COMMON INTEREST AND SUSTAINABILITY

While some environmental reforms may purportedly be in the general interest (everyone presumably wants to halt ozone depletion), the environmental agenda pursued by the centre may not be the same as that which looks most compelling at the periphery. Of course, the centre does not dictate EU policy, though it often presses it forward and the centre is sometimes perceived at the periphery as advancing its own interests through a supposedly universal objective of environmental improvement. If official agencies in the periphery have misgivings about the value and correctness of EU priorities, this is likely to have two consequences. First, it will tend to reduce their willingness to pursue environmental reforms, particularly if those reforms are meeting with domestic opposition. Second, if these criticisms are aired (as is only too likely), this will tend to reduce the respect which is accorded to those EU-supported reforms of which they do approve. The problems here are reminiscent of those surrounding the supposed globality of global environmental problems: if problems are truly global that makes their solution appear urgent, but when North and South cannot agree about which problems are the important ones and which are not, the whole currency of 'globality' is devalued [*Yearley, 1994*]. Any tensions of this sort are likely to be revealed by domestic political rivalries, since it will be in the interest of opposition parties to be seen to be supporting the cause of industry and the local economy against supposedly inappropriate European environmental policies (which may be presented as restrictive or wrong-headed).

The studies in this volume, by throwing particular light on the relationship between environmental protection and economic growth at the periphery, contribute to the current discussion about sustainable development in Europe. As Baker has argued [*1993b*], it is important to acknowledge that the EU's environmental policy is still subordinate to the overall aim of economic growth. Currently, a 'deep green' EU policy orientation, running in the face of growth, is unthinkable. As argued in a deep green article in a recent issue of *The Ecologist* [*Fairlie, 1992: 279*], official support for recycling initiatives is far more hospitable to conventional industry than the policy alternative (re-use) would be. For Fairlie, legislators who congratulate themselves on apparently far-reaching recycling schemes are making much less of a radical difference than they would like to think. In peripheral areas, where demands for economic growth are urgent and where conventional methods for promoting growth (road building, power-plant construction and so on) are readily available, the EU's policy commitment to sustainable development faces a very sharp test.

ACKNOWLEDGEMENT

We would like to express our thanks to Dr. Stephen Young for his highly helpful comments on our work and to all the contributors to the Workshop at which this project was discussed.

NOTES

1. All three authors contributed equally to the writing of this chapter and to the editing of the whole volume.
2. Thanks to the coming into force of the Maastricht Treaty, the European Community (EC) is now the European Union (EU). The new term is used throughout this collection except when its use would be misleading or significantly inappropriate.

REFERENCES

Baker, S., 1993a, 'The impact of recent EC policy on the environment in the west of Ireland', paper presented to the Annual Conference of the Irish Association, Derry, 15–17 October.
Baker, S., 1993b, 'The environmental policy of the European Community: a critical review', *Kent Journal of International Relations*, Vol.7, pp.8–29.
Budd, S. A. and A. Jones, 1989, *The European Community: A Guide to the Maze* (London: Kogan Page).
Campbell, J., 1964, *Honour, Family and Patronage* (Oxford: Oxford University Press).
Croner's Europe, July 1992, Part 2, p.553.
Douglas, M., 1970, *Natural Symbols* (Harmondsworth: Penguin Books).
Douglas, M., 1987, *How Institutions Think* (London: Routledge).
Fairlie, S., 1992, 'Long distance, short life: why big business favours recycling', *The Ecologist*, Vol.22, pp.276–83.
Haigh, N., 1992, 'The European Community and international environmental policy' in A. Hurrell and B. Kingsbury (eds.), *The International Politics of the Environment* (Oxford: Oxford University Press).
Ham, C, and M. Hill, 1985, *The Policy Process in the Modern Capitalist State* (London: Wheatsheaf).
James, P., P. Tayler and M. Thompson, 1987, *Plural Rationalities*, Warwick Papers in Management, No.9, University of Warwick.
Komito, L., 1989, 'Dublin Politics: Symbolic Dimensions of Clientelism', in C. Curtin and T. Wilson (eds.), *Ireland From Below: Social Change and Local Communities* (Galway: Galway University Press).
Lasok, D. and J. W. Bridge, 1989, *Law and Institutions of the European Communities* (London: Butterworths).
Lewanski, R., 1993, 'Environmental policy in Italy: from regions to the EEC, a multiple tier policy game', paper presented at the ECPR Joint Session of Workshops, Leiden, April 1993.
Lipsky, M., 1980, *Street Level Bureaucracy* (New York; Russell Sage).

Majone, G., 1991, 'Regulatory Federalism in the European Community', paper presented at the Annual Meeting of the American Political Studies Association, 29 August to 1 September.
Martin, W. G. (ed.), 1990, *Semiperipheral States in the World Economy* (New York: Greenwood Press).
Mazey, S. and J. Richardson, 1992a, 'British pressure groups in the European Community: the challenge of Brussels', *Parliamentary Affairs*, Vol.45, pp.92–127.
Mazey, S. and J. Richardson, 1992b, 'Environmental Groups and the EC: Challenges and Opportunities', *Environmental Politics*, Vol.1, pp.317–33.
Milton, K., 1991, 'Interpreting environmental policy: a social scientific approach', in R. Churchill, L. Warren and J. Gibson (eds.), *Law, Policy and the Environment* (Oxford: Blackwell).
Nugent, N., 1991, *The Government and Politics of the European Community* (London: Macmillan).
Pressman J. L. and A. Wildavsky, 1979, *Implementation* (California: University of California Press).
Van Schendelen, M.P.C.M., 1993, *National Public and Private EC Lobbying* (Aldershot: Dartmouth)
Siedentopf, H. and J. Ziller, 1988, *Making European Policies Work: Implementation of Community Legislation in the Member-States* (London: Sage).
Skea, J., 1990, *Acid Emissions from Stationary Plant: Reopening the Debate* (London: Friends of the Earth).
Thompson, M. and A. Wildavsky, 1986, 'A Cultural Theory of Information Bias in Organizations', *Journal of Management Studies*, Vol.23, pp.272–86.
Wallerstein, I., 1979, *The Capitalist World Economy* (Cambridge: Cambridge University Press).
Weale, A., 1992, *The New Politics of Pollution* (Manchester: Manchester University Press).
Yearley, S., 1992, 'Green ambivalence about science: legal-rational authority and the scientific legitimation of a social movement', *British Journal of Sociology*, Vol.43, pp.511–32.
Yearley, S., 1994, 'Social movements and environmental change', in M. Reclift and T. Benton (eds.), *Sociology and Global Environmental Change* (London: Routledge).

Policy Co-ordination in Brussels: Environmental and Regional Policy

SONIA MAZEY and JEREMY RICHARDSON

INTRODUCTION

Whereas other contributions to this volume focus upon the *regional* dimension of EU environmental policy, this chapter is centrally concerned with the development and bureaucratic co-ordination of environmental policy at the *European* level. The following discussion highlights EU Commission attempts since the adoption of the 1986 Single European Act (SEA) to integrate EU environmental and regional policies and considers the role of certain environmental groups concerned by the apparent, adverse environmental impact of EU regional policy. EU officials now acknowledge the need for greater co-ordination of these two sectors, not only at the EU level, but also between EU and national authorities which are responsible for implementing policies. Within the European Commission there was little effective policy co-ordination between EU environmental and regional policies prior to 1986. Implementation of the SEA, growing awareness of the contradictory impact of EU regional and environmental programmes, and the salience of environmental issues have since prompted EU organizational and budgetary reforms, designed to facilitate greater co-ordination. These developments are consistent with the central objectives of the Fifth Action Programme which seeks to integrate environmental considerations into other sectoral policies as a means of promoting sustainable development.

The following discussion also highlights the problems of segmented, bureaucratic policy-making within the Union generally. Formally speaking, EU environmental policy is drafted by EU officials within DGXI (Environment, Consumer Protection and Nuclear Safety), whilst EU regional policy remains the responsibility of DGXVI (Regional Policy). In these sectors, EU policy-making at the bureaucratic level is fragmented into specialized fields of competence. As we [*Mazey and Richardson, 1992; 1993*] and others [*Peters, 1992*] have noted elsewhere, policy segmentation is a common characteristic of the EU bureaucracy. The growing importance of these two policy sectors at the EU level since the mid-1970s was not matched by any concerted attempt to co-ordinate the two sectors, at EU or national levels. In consequence, EU environmental objectives have often been either ignored or regarded as being of secondary importance by those EU, national and regional authorities involved in the formulation and implementation of EU regional initiatives. For their part, officials involved in the preparation of EU environmental policy stress that environmental measures must be uniformly applied throughout the Union; no account is or should be taken of peripherality or regional economic conditions in their preparation or application.

Since 1986 attempts have been made at the EU level to ensure that environmental considerations are incorporated into regional developments funded by the Union. In this context, both the SEA (which strengthened the legal basis of EC environmental policy and committed EC member-states to economic and social cohesion) and the concomitant reform of the EC Structural Funds in 1988, were important. The Maastricht Treaty further strengthens the budgetary link between these two sectors by providing for a new Cohesion Fund to help the poorest member-states (Greece, Ireland, Portugal and Spain) meet the costs of environmental and infrastructure developments. The dynamics of these developments are discussed more fully in Section 2. The purpose of the first section is twofold: to highlight the organizational framework and disjointed nature of EU policy-making in these two sectors; and secondly, to outline the new 'governance' mechanisms [*Schmitter, 1990*] which have been introduced by the European Commission, as a means of fostering closer integration between these two sectors. The picture which emerges is one of incremental development. Whilst the SEA initiated important formal changes, internalization of these changes is still underway.

1. FROM SECTORAL TO INTEGRATED POLICY-MAKING

1.1. *The Development of EU Environmental Policy*

The development of EU environmental policy need not be rehearsed here since this information has already been provided elsewhere (see Introduction). The Fifth Action Programme, adopted in March 1992, is of particular significance to this particular article in the sense that is the first to adopt a 'holistic' approach to the environment. In short, the central objective of the current environmental action programme (EAP) is the promotion of 'sustainable development', which is to be achieved by the incorporation of environmental considerations into other economic and sectoral policies [*Judge, 1992*]. Within the Commission, the growing importance of the environment resulted in the establishment, in 1981, of DGXI, which has primary responsibility for ensuring the preparation and implementation of EU laws. However, several other departments have responsibility for specific aspects of environmental policy. Research-related aspects of EU environmental policy are the responsibility of DGXII, DGIII (Internal Market) is responsible for the implementational aspects of several environmental directives, whilst both DGVI (Agriculture) and DGVII (Transport) contain specialized departments dealing with environmental and ecological questions. Other aspects of environmental policy are the responsibility of DGs V (Employment, Industrial Relations and Social Affairs), XIV (Fisheries), XVII (Energy), XVIII (Loans and Investments) and the European Investment Bank. Meanwhile, the European Environment Agency (EEA), the creation of which was agreed by the Council on 7 May 1990, will provide technical and scientific advice to the Commission and EU member-states.

DGXI has in the past been (and in some cases still is) regarded with suspicion by some producer groups, national administrations and other Directorates-General within the Commission. This tension is due partly to the close historical links between DGXI and environmental lobbies such as the European Environmental Bureau (EEB), World Wildlife Fund (WWF) for Nature, and the Institute for European Environmental Policy (IEEP). As we have argued elsewhere, one obstacle to more effective policy co-ordination in this sector is the (to some degree, self-fulfilling) belief that DGXI is a 'captured' agency [*Mazey and Richardson, 1992*]. A related problem is the apparent conflict between the Union's commitment to the promotion of the market-led growth (epitomized by the 1992 programme) and environmentalism. Though recent developments suggest that both sides in this debate now acknowledge the potential complementarity of these goals, this has not always been the case [*Weale and Williams, 1992*]. The belief (however misguided) that DGXI was essentially unsympathetic to the interests of industry and commerce has in the past been an important barrier to horizontal co-ordination within the bureaucracy.

The SEA was undoubtedly a turning point in the *formal* development of EC environmental policy. Articles 130R and 130T of Title 7 of the Act are devoted to environmental concerns, thereby incorporating the sector into the EC Treaties for the first time. Significantly, these Articles make it a general requirement that environmental concerns should be integrated with other EC policies. They go on to state that the object of EC environmental policy is to preserve and improve the quality of the environment, protect human health, and encourage the rational utilization of natural resources. The SEA also specifies that Community action is to be based upon three guiding principles: preventive action should be taken; priority should be given to making good environmental damage at source; and the polluter pays principle [*Commission of the EC, 1990a*]. These changes provided the necessary legal basis for the subsequent integration of environmental policy with other sectors, including regional policy. The budgetary link between these two sectors was further strengthened by the December 1992 Edinburgh European Council decision to create a new Cohesion Fund to help the poorer member-states (Greece, Ireland, Portugal and Spain) meet the costs of environmental and infrastructure developments [*Commission of the EC, 1993a*].

In reality, the Union's environmental expenditure is spread across various budgetary appropriations, and there is little doubt that this has also served to impede effective policy co-ordination. Nevertheless, it is clear that overall environmental expenditure levels have begun to rise quite steeply. In January 1991, for instance, the Commission estimated that, as from 1989, environment-related expenditure ought to amount to an average of approximately 650 million ECU per annum, compared to 135 million ECU in previous financial years [*Court of Auditors, 1992*]. Around 1.2 billion ECU of this was earmarked for projects for the period 1989–93 in the regions identified as being less developed [*Commission of the EC, 1990a*]. More recently, at the Edinburgh Summit in December 1992, EC heads of government agreed to double environment-related Structural Fund expenditure in the next financial perspective (to approximately seven billion ECU). An additional 2.6 billion ECU was allocated to the Cohesion Fund.

1.2. *The Regional Dimension*

The Structural Funds – the European Regional Development Fund (established in 1975), the non-price support section of the European Agricultural Fund and the European Social Fund (created in 1961) – constitute the principal financial instruments of EU environmental policy. Since the relaunching of the internal market programme in the mid-1980s these funds have been substantially increased, from 7000 million ECU (19 per cent of the EC budget) in 1987 to 14,000 million ECU (25 per cent of the EC budget) in 1993. Furthermore, the rules governing their allocation changed in an attempt to reduce overall economic disparities within the Union and promote the least favoured regions.[1] In accordance with the new procedures, the bulk of the funds are now allocated to multi-annual (beginning 1989–93) programmes known as Community Support Frameworks (CSFs), deemed by the EU Commission to be compatible with the five funding Objectives agreed by EU member-states.[2] The introduction in 1988 of a 'structural policy' based upon the three Funds listed above necessitated greater institutional co-ordination. Thus, a new Directorate General, DGXXII (Co-ordination of Structural Instruments) was established to co-ordinate the three Structural Funds and their interventions with those of other EC agencies such as the European Investment Bank [*Marks, 1992*].

The shift away from individual project funding to the financing of multi-faceted CSFs was initiated by the Commission in 1984 with the development of the Integrated Mediterranean Programme (IMP) and was extended in the 1988 reforms. This change was introduced as a means of increasing Commission influence over the allocation of the Funds and promoting integrated, medium-term regional development. In addition to the CSFs, the Commission allocated a further 3.8 billion ECU to other Community Initiatives (CIPs) for the period 1989-93. These include: VALOREN (grants for exploiting indigenous energy sources); RECHAR (conversion of coal-mining areas); ENVIREG (protection of the environment in lagging regions); LEADER (fostering of innovatory approaches to rural development); RENAVEL (conversion of shipbuilding areas); RESIDER (conversion of steel areas); REGEN (integration of gas and electricity transmission networks); PRIMA (assistance for lagging regions to meet EU-wide quality standards); and REGIS (to assist integration of the most remote [overseas] regions). Clearly, all of these also have important environmental implications. Indeed, the ENVIREG programme was set up in May 1990 specifically to link the environment to regional development in coastal areas, particularly in the Mediterranean region [*Commission of the EC, 1990*].

Just as in the environmental sector, implementation of the SEA prompted a strengthening of EU regional policy. The 1988 Structural Fund reforms also introduced two new guiding principles of EU regional policy: additionality and partnership, both of which have been influential in extending the relevant policy network to include sub-national authorities. Additionality means that EU funds must be used in conjunction with (not instead of) national assistance. The partnership principle reflects the Commission's claimed commitment to subsidiarity and its desire to involve diverse actors and organizations at all levels, local, regional, national, public and private, at each stage in the Funds' operation [*Commission of the EC, 1991*]. As discussed in the following section, environmental groups have been quick to claim their right to be included in this list.

1.3. *'Never the Twain shall Meet'? Integrating Environmental and Regional Policies at the EU Level.*

As indicated above, environmental and regional policies have both become essential concerns of the EU Commission. This development is reflected in the growing financial and institutional importance of these sectors at the Union level. Since 1986 the European Commission has sought to establish a more coherent approach to policy-formulation and implementation in both fields. This has entailed the formal acknowledgement of the need for closer co-ordination between these sectors. In budgetary terms, the links are already strong since the Structural Funds constitute the principal financial instruments of the environmental action programme.

In reality, however, regional initiatives funded by the Union have in the past often been at odds with EU environmental objectives. Growing awareness of this problem prompted an investigation into EC environmental action by the Court of Auditors in 1990-1. The resulting Report was highly critical of the poor co-ordination and monitoring of environmental initiatives which, as the authors stressed, had often served to limit and even undermine the impact of Community action in this area [*Court of Auditors, 1992*]. In the aquaculture and fish-farming sectors, for instance, funding was provided for several intensive exploitation projects, notably in the lagoons and other wetland areas along the Mediterranean, and charged to appropriations earmarked for fisheries and additional IMP or ERDF appropriations. Simultaneously, DGXI was providing financial aid to environmental protection agencies wishing to purchase lagoons in order to protect them from intensive exploitation. Nor was this problem confined to the European level. In various cases examined by the Court, there was little co-ordination between national government departments concerned with environment-related matters. Lack of space precludes a full discussion of this Report; the following paragraphs thus concentrate on those sections of the Report dealing with the environmental impact of EC-financed regional programmes.

The Court was particularly critical of the lack of consideration given to the environmental impact of EC structural policies by EC, national and regional authorities. It attributed this to the fragmented nature of the policy-making processes (at all levels) and the inadequate environmental monitoring of Structural Fund expenditure by the Commission. The Court's checks confirmed that reform of the Funds had not prompted any appreciable progress at the national level in procedures designed to give more attention to environmental problems in the drafting of regional programmes. These were prepared by the departments responsible for regional development, agriculture and employment, each in its own field. In most cases, departments responsible for the environment were not consulted and in many instances regional and national authorities were ignorant of EC environmental provisions. In various regions the Court itself had to supply departments responsible for the preparation of regional programmes with details of Community law regarding the protection of the environment.

Secondly, with regard to the content of regional operations presented as being beneficial for the environment (and specifically funded for this purpose), the Commission discovered a bewildering array of initiatives, not all of which were obviously compatible with the objectives of the Fifth Action Programme. Thus, notwithstanding EC concern over the adverse environmental impact of urbanization in coastal regions, the operational programme for a region in Southern Italy provided for the construction of medium and large hotels up to a capacity of 1000 beds around a bay considered to have tourist potential. Moreover, the monitoring team found that several programmes financed by the Structural Funds actually contributed to the deterioration of the environment. For example, work carried out on the port of Salonica (financed within the framework of the IMP) entailed deepening one of the docks and draining the surrounding area. The three to four million cubic metres of excavated, polluted mud was poured into the bay, thus contributing to the destruction of already badly damaged flora and fauna [*Court of Auditors, 1992*].

A third problem highlighted by the Report was the widespread absence of an integrated approach to environmental policy. Regional authorities concerned tended to stress the need for piecemeal investments aimed at reducing pollution rather than adopt a 'holistic' approach which would ensure that the measures in question do not contribute to the destruction of the environment. The construction of the Athens metro, for instance, had been supported by the Community (178 million ECU) as an environmental project, yet no attempt had been made by the regional authority to integrate this into an urban traffic control policy. Separation of responsibilities between environmentally oriented departments responsible for running the ENVIREG programme and those public works departments which were running other infrastructure programmes, was highlighted as a major obstacle to the development of such a global approach. Numerous water purification stations, for instance, have been built in rural villages and towns with Structural Fund support. However, many of the stations are non-operational because local authorities have insufficient funds to maintain them. Moreover, in some member-states, environmental policies such as pollution control are entrusted to departments responsible for hygiene and health which have other higher-priority tasks and which have not been specially adapted to perform complex environmental functions such as 'environmental audits' [*Court of Auditors, 1992*].

Fourthly, the Court discovered that in some cases, EC environmental clauses in EC-funded programmes were quite simply not applied, either because contractors were unaware of them, or in other cases, because it was their opinion that the relevant national laws were sufficient. Similarly, the Court discovered that in some regions water and air quality checks were not systematically carried out. Thus, the Report concluded: 'There is not, therefore in the measures receiving aid from the Structural Funds, any very clear link between the investments financed and the protection of the environment. In those cases where measures of this sort are actually adopted, there is little follow up and their real impact is rarely monitored' [*Court of Auditors, 1992*].

1.4. *Towards More Effective Policy Co-ordination*

The central point to emerge from the above Report was that EC environmental objectives cannot be met by action at the EC level alone, but only on the basis of a shared response, involving a sharing of responsibility across administrative sectors and at all levels. This point has now been acknowledged within the Commission, notably within DGXVI, which now accepts the need to incorporate environmental objectives into the preparation of CSFs financed by the Structural Funds. Recognition of this fact has already prompted a number of organizational changes at the EU level, designed to increase the transparency and administrative co-ordination of environmental policy. DGXVI, for example, has begun to develop more effective policy evaluation techniques. It has also set up working groups to consider the relationship between jobs and the environment and to develop environmental indicators.

Specific reforms address the problem of horizontal and vertical co-ordination. Thus, all Directorates-General have been asked to calculate aggregate totals for all spending on environmental measures and send DGXI a memorandum containing the results of this exercise. The Commission is also currently preparing a system for the statistical collection and presentation of environmental indicators and itemized EU environmental expenditure. Secondly, the Commission has established detailed management plans at both Commission and DG level which seek to plan and prioritize environmental activities, to relate these initiatives to key target areas such as energy and agriculture, and to define performance targets, deadlines and so on. Thirdly, the Commission has increased the amount of resources available for inter-departmental co-ordination between Commission departments with responsibility for environmental policy. Initiatives which are intended to strengthen vertical co-ordination include DGXI's recommendation that member-states be required to submit to it an 'environmental profile', depicting the state of the nation's environment and national policy priorities in this sector. There are also plans to establish an Environmental Policy Review Group comprising representatives of the Commission and the member-states at Director-General level to exchange views on environment policy. The creation of a network comprising representatives of the relevant national authorities and representatives of DGXI is also envisaged to ensure the effective implementation of EU measures [*Court of Auditors, 1992*].

Attempts are also now being made to co-ordinate the numerous sources of EU environmental expenditure. With regard to the Structural Funds, the Commission believes that the principle of subsidiarity means that member-states must be encouraged to involve their environmental authorities (and possibly non-governmental organizations) more closely in the definition, costing and implementation of regional programmes. In the case of operational programmes, member-states are now required to inform the Commission in advance of arrangements made to monitor the environmental impact of the planned measures. All applications for Structural Fund support or EIB loans are also scrutinized by DGXI in order to ensure that the proposed operations are consistent with EU environmental objectives and include sufficient environmental expenditure appropriations. Proposals which fail to meet these criteria are rejected whilst ongoing projects deemed to be unsatisfactory in this respect may have their EU funding blocked. DGXI has also begun to centralize its finance and contract functions and is in the process of creating an integrated financial data-base to monitor contracts and programmes more effectively. A new financial instrument, LIFE, was also created in December 1991 as a means of strengthening and increasing the effectiveness of organizational structures, controlling pollution, and protecting sensitive areas. This fund is also controlled by DGXI.

Many of the reforms listed above emerged from the Commission's reply to the Court's report and are currently in the process of being implemented. It is, therefore, far too early to assess their impact upon EU environmental policy. The overall objectives of these changes are, however, clear. The effective implementation of the Fifth Action Programme requires the integration of environmental concerns into other sectoral policies. This can only be achieved by effective administrative and financial co-ordination. Strengthening DGXI's co-ordinating role in these respects whilst at the same time devolving responsibility for the operational management of programmes may be the most practicable means of proceeding, but *Commission* level co-ordination will also be needed.

2. SECTORAL EROSION: ISSUE SALIENCE, POLITICAL MOBILIZATION AND BUREAUCRATIC POLITICS

The history of EU environmental policy to date has exhibited the same characteristics of EU policy-making more generally, namely a high degree of segmentation, though integration has begun to develop between the two policy sectors under discussion. These are the beginnings of what may be a quantum change in inter-sectoral co-ordination, although, as we argue later, the actual implementation of these changes still presents significant administrative and political challenges. Leaving aside this issue, how do we explain the dynamics of this process of sectoral erosion? Below, we present a number of possible explanations, though further research is needed for us to be able to scale or weight the various factors which appear to have been at work.

2.1 *Issue Salience*

In what is still the classic account of the process through which issues pass over time, Anthony Downs suggests that one of the phases of the so-called 'issue attention cycle' is when decision-makers and public alike have 'discovered' a problem. A kind of euphoria then develops concerning the need for and possibility of finding solutions [*Downs, 1973*]. In a similar way, Gregory noted that 'the environment' in the 1970s developed a 'halo effect', which (temporarily he feared) would enable those actors pressing for environmental improvements to exercise greater influence in the policy process [*Gregory, 1971*]. Both writers might be surprised that some twenty years later, the 'environment' has regained the degree of issue salience that gave it such dominance over most other policy problems in the 1970s. Whatever the causes of this resurgence (no doubt Chernobyl and the continued reporting of the worsening effects of ozone depletion and global warming are important), there is little doubt that 'environmentalism' is back in centre stage of political debate. Eurobarometer data consistently confirm that respondents continue to rate environmental issues as a very high priority. For example some 85 per cent of respondents in the Eurobarometer survey (reported in July 1992) regarded environmental protection as 'immediate and urgent'. This figure reflects a rise of 11 per cent over the previous year.

More importantly, key interests have concluded that the environment as an issue is here to stay and may have become institutionalized. For example, many multi-national companies, such as ICI, have begun to integrate environmental concerns into their planning processes. Similarly, BT (British Telecom), which was generally thought to have been a rather 'clean' industry, has 'internalized' the issue. The Deputy Chairman of BT recently reported that 'some time ago, we in BT projected that all companies, in all industrial sectors, would soon be expected to address the environment as a serious business issue and, more especially, would need to demonstrate to their shareholders that they were doing something positive about it' [*The Independent on Sunday, 14 March 1993*]. Indeed, in 1993 BT won an environmental reporting award from the Chartered Association of Certified Accountants for a report, which on BT's own admission, shattered the public delusion that it was a 'clean' company. The Report prompted changes in BT's internal accounting systems, designed to take account of environmental problems. These reforms put the company ahead of competitors in the marketplace where environmental factors are becoming a common feature of industrial procurement policies. The accountancy profession itself had earlier presented a report to the government by the profession's advisory committee on business and the environment. Thus, some key interests (namely polluters) have begun to shift their attitude on environmental issues. Simultaneously, many environmental groups have begun to develop more 'responsible' lobbying techniques and exhibit less ideological or less 'pure' positions on specific environmental issues. In a very real sense, not only has 'the environment' regained its high political salience, but the constellation of interests surrounding the issue has shifted to more manageable centre ground, where consensus and compromise are possible. This in itself makes policy co-ordination more possible.

Majone's suggestion that intellectual and cognitive factors play a crucial role in the shaping of policies is clearly illustrated in this case study [*Majone, 1989*]. The high political salience of environmental issues has forced a range of policy-makers to address 'environmentalism' within their own policy sectors and organizations, even when their primary focus is quite different. Thus, environmental concerns have increasingly become internalized in the cognitive and intellectual processes of disparate institutions, including DG XVI and its 'clients', (regional/local economic and politico-administrative interests). In short, high political salience is itself a co-ordinating factor – it is the 'spark' which forces disparate (and often conflicting) organizations to address the same problem. In the Downsian 'pre-problem' phase, these inherently cross-sectoral problems tend to be identified only by those policy professionals at the sectoral, or more usually, sub-sectoral level. As these problems achieve greater political salience, other sectors and sub-sectors begin to pay attention, and may even jump on a bandwagon, or at least piggy-back their own concerns onto the new issue. If an issue achieves sufficient political salience, it then becomes as much an opportunity as a threat to many policy-makers. They can re-package old programmes, and justify new ones in the light of the new issue, and even gain extra resources as a result. At a certain point in the trajectory of an issue, it becomes rational to take it on board rather than to resist it.

2.2. *Political Mobilization*

It may be a truism that modern politics is really about organization, that issues are organized into or out of politics, but it is an important observation nevertheless. In this case, organizations, particularly certain environmental and voluntary organisations, which are increasingly active at the European level, have been adept in exploiting the high salience of the environment to penetrate new policy sectors. In so doing they have themselves constituted an important co-ordinating factor. The EU-level campaign strategies of the World Wide Fund for Nature (WWF) are discussed below to illustrate the way in which environmental groups generally have maintained and, by their action, increased pressure upon EU (and national) policy-makers to take account of the environment. The WWF contacted DGXVI in 1988, because the organization was concerned by the apparent lack of mechanisms for assessing the environmental impact of the reformed Structural Funds. In its view, this problem had been exacerbated by the shift to a programmatic approach by DG XVI. Basically the pace of reform and the novelty of the CSFs had left DGXVI without appropriate evaluation criteria. In this sense, DGXVI fitted the classic model of a bureaucracy charged with 'solving' an urgent problem. Precisely because of the need to implement the reforms quickly, evaluation mechanisms had not been incorporated at the policy design phase. They became an issue only when the adverse effects of the policies came to be questioned by other interests and institutions.

Ironically it was the moves towards greater co-ordination within the various EU funds affecting disadvantaged regions which appears to have been the 'spark' for the now intense lobbying that the reform process has attracted. A WWF/IEEP Briefing on the subject of EC Structural Fund reform, published in 1989, stressed that '. . . it must be recognized that the injection of large sums of development money into predominantly rural regions, often with fragile local economies must be planned and executed with considerable sensitivity. Many of the regions concerned have delicate environments which are easily damaged or even destroyed by misplaced forms of development' [*WWF, 1989: 3*].

Citing the earlier example of the EC's IMPs, the Briefing argued that it was of 'utmost importance' that these problems should not be repeated in the course of much larger scale operations to be financed by the Structural Funds. Pressure was also applied to the Community to honour the decision taken at the Rhodes Council in December 1988, that is, that member-states integrate environmental protection into other Community policies and make sustainable development an over-riding objective.

For WWF there were two main issues surrounding the 1988–89 reform of the Structural Funds. The first concerned the need to ensure that environmental considerations were incorporated into funding applications submitted by the member-states. Secondly, WWF was anxious that Structural Fund allocations should permit environmentally sensitive development and assist badly needed conservation initiatives [*WWF,*

1989: 7–8]. The WWF welcomed new rules (the so-called internal instructions), which from 1 April 1989 required national governments to state how the environment was being taken into account in all plans. However, the WWF Briefing added that '. . . it is important that independent outsiders and NGOs are able to have access to plans and programmes submitted, to see whether they do contain information about the environment and to see whether it is accurate' [*WWF, 1989: 7–8*]. This interpretation of the new rules represented an important challenge to the power of national governments. Moreover, WWF also argued for an increase in the EC's bureaucracy by suggesting that DGXI should be expanded and given the resources to create a fully staffed and qualified unit to deal with the volume of development proposals that the Structural Funds would generate a nice example of the type of agency/client relationship which Downs has identified as a characteristic of many bureaucracies! [*Downs, 1967*]. Finally, it is important to note in the context of this discussion that the WWF Briefing was designed '. . . to alert environmental organisations and others to what has been agreed, and what it might mean for the environment' [*WWF, 1989: 7–8*]. In other words, it was intended to stimulate other national, local, and regional environmental organizations to become involved in the new regional and investment programmes which were being drawn up in 1989.

In 1989 the Commission's own 'Task Force' acknowledged that Commission procedures for ensuring the integration of environmental considerations into Structural Fund appropriations were inadequate [*Commission of the EC, 1989*]. Identification of this problem had resulted in environmental protection measures being incorporated into agreements between member-states and the EU on Structural Fund allocations after 1989. This was, WWF and IEEP conceded, 'a very significant advance on the position before the reform of the Funds when environmental considerations played little part in the negotiations and were rarely referred to in key documents' [*WWF/IEEP, 1990: 2*]. Nevertheless, the same WWF/IEEP publication then proceeded to list a catalogue of remaining problems, similar to those highlighted by the Court of Auditors' Report discussed above [*Court of Auditors, 1992*]. Both groups stressed the need for greater public access to the Operational Programmes, (most of the CSFs had by then been approved) and continued to campaign for a larger and better funded environmental scrutiny and support unit within the Commission. Reflecting on the often close links between environmental groups and agencies, they also pressed for environmental ministers and agencies to participate in the monitoring process and for environmental ministry representatives or their nominees to be included in all steering committees. They also demanded that effective consultation procedures to be established with NGOs [*WWF/ IEEP, 1990: 8*].

Essentially, the argument was about the groups' *rights to participate* in the policy formulation and implementation processes regarding the Structural Funds. This reflected a belief that the composition of policy communities and networks (in our terminology) is important in influencing policy outcomes. Prior to the more recent reforms outlined above, participation was certainly restricted and environmentalists (including even environmental agencies) were not included in any list of who really 'mattered' in terms of the formulation and implementation of detailed proposals.

In a key passage, the WWF/IEEP argued that:

> The underlying need is to recognize the environment as fundamental to regional and economic planning and assistance, rather than as a factor only to be considered in marginal cases or where development affects an especially sensitive site. Both trade unions and industry are recognized as legitimate 'social' partners in guiding the direction of certain strands of economic development. Environmental agencies too, both public and private, should be drawn more closely into the planning process rather than left to react in the wake of ill-conceived or insensitive projects. Under the new arrangements for the Funds, there is a welcome intention to involve local and regional authorities much more closely in the planning of regional aid. This must be respected and broadened to include environmental agencies. Wider participation is one of the foundations on which greater sensitivity to the environment must be built [*WWF/IEEP, 1990: 15*].

The 1992 review of the Structural Funds, provided a timely opportunity for environmental groups and various environmental agencies (including DGXI) with whom they had links. In the preceding months the network of actors concerned with EC environmental policy was mobilized and pressure increased for yet further integration of the two policy sectors. Thus in February 1992, 58 national and seven international environmental groups in Europe published a collective statement concerning the impending reform of the EC Structural Fund Regulations. A significant proportion of the signatories were from peripheral regions, for example, four from Ireland, five from Portugal, six from Greece and no fewer than 14 from Spain. In fact in one region of Spain alone, Castilla-la-Mancha, it was estimated that there had been 27 infringements of EC, national and regional environmental laws in the preceding three years as a result of projects funded by the Structural Funds.

The environmental lobby was performing a role to which it has become well suited: forcing issues onto the political agenda, often via their ability to monitor and expose implementation failure. Since 1988, the environmental network had been monitoring the implementation of the reformed Structural Funds. Its own reform agenda was by now familiar: widening the policy network to include environmental groups and pressing the Community to implement the commitment to sustainable development it had undertaken in Rhodes. A list of five specific amendments were proposed to the Framework legislation (EEC 2052/88) and the Co-ordinating Regulation (EEC 4253/80) due for revision in 1992. This list included an amendment which, if adopted, might benefit the environmental groups themselves, that is, that Community expenditure be permitted for technical and managerial support to ensure that Structural Fund assisted projects achieve environmental sustainability [*WWF, 1992a: para. 15*]. As we suggest below, this proposal opens up a completely new strategy for these groups to become more directly involved in the preparation and application of proposals.

As is typically the case in the EU policy process, at a certain stage in the life of an issue, an opportunity is found to bring the key actors together in some kind of *ad hoc* forum. Such a meeting took place in Brussels on 22 May 1992, (funded in part by DGXI). This so-called 'experts meeting' brought together Commission officials from various DGs with an interest in the incorporation of sustainable development principles into EU Regional Policy. Considerable agreement emerged from this meeting (a good example of the creation of an identifiable 'policy community' within which fellow professionals addressed a common problem). As far as the environmental lobby was concerned, three issues were now at stake: sustainable development; involvement of various environmental actors in the planning process, that is, widening the basis of 'partnership'; and transparency [*Letter from WWF Director, 2 June 1992*]. Not only were these objectives transmitted to the Director Generals (and some of their staffs) in DG XXII, XVI, XI, VI, and V, they were also communicated to other EC institutions, member-state governments, regional and local authorities and other parties in the network of over 60 environmental organizations throughout Europe supporting the campaign.

Two subsequent opportunities have arisen to maintain what must have seemed to Commission officials and MEPs a cacophony of environmental lobbyists. First, the Danish Presidency ensured that this issue remained high on the Community's agenda. As we have argued elsewhere, groups can use the Presidency to press particular issues [*Mazey and Richardson, 1993*] and this was precisely what occurred in this case. It was apparently the Danish WWF which informed the Danish Environmental minister of the problem and persuaded him to press for further co-ordination and integration of environment and regional policy, during the Danish Presidency. In a *Briefing for the Danish Presidency* the 13 WWF National Organizations in Europe (eight EU and four EFTA) stated 'that one example where the gap between the intention of the Treaty and practical reality is in danger of widening concerns the requirement for the integration of the environment into the European Community's other policy areas . . . highlighted by the Court of Auditors Report in 1992' [*WWF, 1992b*]. Yet again the general argument, supported by many other NGOs such as the European Citizens' Action network, for wider participation, greater access and so on, was highlighted in respect of Structural and Cohesion Funds.

2.3. *Judicial Politics*

Environmental groups which do not generally enjoy the degree of access to policy-makers and resources of, say, multinational corporations or industrial and commercial trade associations, have also resorted to legal action before the European Court of Justice. The now infamous case of Mullaghmore in Ireland is the subject of an ECJ case, lodged on 4 December 1992 by An Taisce, The National Trust for Ireland and WWF UK. The case reflects environmentalists' concern over the interpretation which the Commission places on its duty under article 130R of the Treaty and Article 7.1 of the Structural Fund Regulations (2052/88) when considering whether particular projects should be part-aided by the Commission. In this case, the Commission is alleged to have failed to take into account the objectives of both the Habitats Directive (92/43/EEC) and the Berne Convention (Convention on the Conservation of European Wildlife and National Habitats) to which the EU is a party in its own right. The pending case constitutes yet a further source of pressure on the Commission generally, both at Directorate General and Secretary General levels, to introduce the desired changes in the draft regulations concerning the operation of the Structural and Cohesion Fund programmes.

More generally, the case raises a fundamental issue in terms of policy co-ordination between these two sectors, namely the precise legal definition of EU environmental policy. Throughout this whole process, the European Parliament (EP) has been quite active via the Environment, Public Health and Consumer Protection Committee, chaired by Ken Collins MEP. Indeed, the influential Court of Auditors' investigation referred to earlier was itself a response to Parliamentary pressure. The EP (Environment) Committee, took up the implications of the Irish case; in January 1993 it tabled an Oral Question With Debate (under rule 58 of the Rules of Procedure) for the Commission. In reply to earlier questions

(H-1051/92 and H 1135/92) the Commission had stated that neither the Berne convention nor the Habitats Directive (cited in the above ECJ case) formed part of EC environmental policy. Yet, Article 7(1) of Regulation 2052/88 requires all Structural Fund activities to comply with Community legislation and Community policy. The Committee therefore asked the Commission to define Community Environmental policy for the purposes of the above Regulation. In particular, it requested the Commission to specify whether or not the policy encompassed other EC environmental instruments such as: the Berne Convention; the Habitats Directive; the Biodiversity Convention; the Fifth Action Programme; the Bonn Convention on the conservation of migratory species of wild animals; and the Council's decision to reduce CO_2 emissions to 1990 levels by the year 2000. The Committee also asked the Commission to explain what internal procedures it used to ensure that the requirements of Article 7(1) were met.

This move has further increased the pressure on the Commission to deliver more effective policy co-ordination. The pressure is especially hard on DGXVI, but it is by no means confined to it. In terms of interest group lobbying, for example, WWF Europe devotes at least 50 per cent of its lobbying effort to trying to influence DGs other than its 'sponsor', DGXI, and now has contacts across a wide range of the Commission's activities.

2.4. *Bureaucratic Politics*

As Peters has observed, a key feature of the EU policy process is bureaucratic politics [*Peters, 1992*]. Predictably, the Commission has itself taken steps to manage this potentially damaging policy conflict. It is in nobody's interest within the bureaucracy that the issue should be processed exclusively in the public area, not least because policy outcomes are highly uncertain in such cases. Stability and some degree of predictability need to be restored if the lives of officials are not to be made intolerable! The destabilization of the operating environment of DGXVI, in particular, has produced a more 'mixed' style of policy formulation. Organizational cultures rarely change overnight and there seems little doubt that DGXVI, as with all other DGs, will continue to practise much internalized policy formulation. Nevertheless, as already indicated, DGXVI has taken on board a number of the issues raised by the EP, the Court of Auditors, DGXI, and environmental and citizens' action groups. To some degree, there appears to be a change in organizational culture within at least parts of DGXVI, with some recognition that the environmental industry and environmental protection has some potential to be an economic engine in terms of regional development. Thus, DGXVI is beginning to utilize the environmental issue, rather than resisting it and defending its policy space. It has, for instance, agreed to amend some Regulations in order to take greater account of environmental regulations, (though the Commissioner is still insistent that there should not be higher environmental standards for the peripheral regions than elsewhere in the EU).

Just how far the policy process will produce policies which are more environmentally sensitive in practice remains to be seen. The evidence so far suggests that the process of institutionalizing and internalizing environmental questions is well underway within the Commission, just as it is inside private sector organizations such as multinational firms. Both the degree of political salience of the issue and its longevity now make it rational for policy actors, public and private alike, to 'accommodate' the issue in their policy planning processes. Thus, in terms of bureaucratic politics, environmentalism has begun to contribute significantly to the erosion of administrative boundaries within the Commission and to a more integrative policy style. For example, in June 1993 the European Commission adopted an internal communication outlining how it intends to achieve better integration of environmental protection requirements in other policy sectors [*Commission of the EC, 1993a*]. The document commits officials to:

(i) consider the environmental implications of all actions and where these are 'significant' undertake an environmental impact assessment;
(ii) describe, justify and consider the environmental costs and benefits of legislative proposals with a 'significant' environmental impact;
(iii) regularly examine the progress made with regard to the integration of environmental considerations into Community policies. (This will be carried out on the basis of an evaluation by each DG of its own environmental performance);
(iv) nominate a senior official in each DG to ensure that legislative proposals take account of the environment and the need to contribute towards sustainable patterns of development;
(v) set up a special unit in the DG for the Environment to co-ordinate and implement the Action programme;
(vi) prepare a code of conduct on the Commission's own operations from an environmental point of view, notably relating to purchasing policies, waste prevention and disposal, energy saving;
(vii) publicize information on developments and progress in this area through the Work Programme and in the Commission's Annual Report of its Activities [*Commission of the EC, 1993b*].

Meanwhile, the behaviour of groups, including regional and local authorities, has both contributed to and been affected by this process. Certain types of groups, particularly the environmental lobby and other NGOs, have been very effective in both maintaining issue salience and in utilizing this salience to gain access to 'new' (for them) policy sectors. They have also pressed for procedural changes which might ensure continued access in the future. But the groups are also themselves in turn being affected by this process of change. Just as sectoral boundaries are being eroded, so are those of existing policy communities and networks. This means that policy-making is likely to be a more widely shared process, involving the building of new and rather more complicated coalitions. For example, DGXI has recently begun to have much more contact with regional and local authorities. This reflects a recognition within DGXI that in order to be effective it has to exert greater influence on implementation-processes which are often not its responsibility – approximately 40 per cent of the Fifth Action Programme will be implemented by regional and local authorities in the EU. It also reflects a recognition by regional/local authorities that Structural and Cohesion Fund assistance will in future be dependent upon the accommodation of environmental protection measures. Meanwhile, DGXVI has begun to extend its traditional 'client' group to include the environmental groups. Thus, policy co-ordination is being encouraged by an increasing amount of client 'sharing' between DGs within the Commission generally, and particularly between DGXI and DGXVI. This is exposing policy-makers to broader political pressures, but, more importantly, it is expanding the market for policy ideas within the respective DGs, ultimately the most effective form of policy co-ordination.

3. CONCLUSION: THE DEVIL IS IN THE DETAIL

The main thrust of this paper is not that co-ordination problems have been solved, but that a process of change is taking place at the centre. This does not mean that DGXVI or indeed other key DGs such as DGV are now 'green'. Our argument is that there has been a degree of greening of DGXVI, as there has been of many organizations in Western Europe and elsewhere. No doubt there are still many in DGXVI who merely pay lip-service to the need to integrate environmental concerns into regional policy. The objective of the DG has not changed: it is concerned primarily with economic regeneration. The current (1993) Commissioner, Bruce Millan, is likely to maintain a stout defence of these objectives. Moreover, as most implementation theorists argue, the power enjoyed by implementors is typically very considerable. The EU is plagued by implementation failure and we should not expect the policy area under consideration suddenly to become a marked exception. No-one who has read the Court of Auditors' Report could underestimate the enormous difficulties in applying Structural Fund and Cohesion policies in an environmentally sound way. We have described the beginnings of change, not the successful implementation of change.

One of the main issues presented by the environmental and citizen action movement – wider participation on the planning process – has already been conceded. The draft regulations now include a statutory widening of the consultation process to include environmental authorities at the national, regional and local levels. This means that much broader policy considerations are likely to be brought to bear in the formulation of funding proposals and in the planning of schemes. Although NGOs have not been given statutory consultation rights, more information will be available and they will be able to utilize their often almost symbiotic relationship with some official environmental agencies. Problems may persist, however, in those countries where 'environmental' tasks are performed by other authorities such as Departments of Public Health and/or where 'environmental' authorities are primarily responsible for other tasks such as road building! Moreover, as suggested above, regional and local authorities are beginning to realize that EU-funded regional development will be jeopardized if the environmental issue is not addressed.

The political need, on the part of DGXVI to be seen to be more environmentally aware, and the need for regional and local authorities to minimize the risk of rejection of funding applications or legal challenges, has presented environmental groups like WWF with an opportunity to exert direct influence on the implementation process. Thus, WWF and others now have the possibility of presenting themselves as 'problem solvers' to both DGXVI and regional/local authorities. Adopting a new style of 'lobbying', WWF is becoming very active at national government level and plans to target regional authorities. It also plans to organize meetings at the regional level at which it will make presentations, lasting up to one and a half days, in order to focus public attention on the ways in which conservation, rural development and job creation can be rendered mutually compatible. As WWF already has field experience (sometimes funded by the EU) it is relatively well-placed to play a consultancy role to the regions and localities, in both demonstrating the viability of a more environmentally conscious development strategy and assisting the authorities to secure EU funding. Equally, WWF hopes to secure technical assistance money from the Structural Funds (as is now possible).

If this WWF strategy works, it will mark a significant departure in the organization's lobbying strategy, which is not, however, without risk to WWF. If it is involved in assisting regional/local authorities to formulate funding proposals and also in receipt of technical assistance funding from DGXVI, it risks being associated with the failures as well as the successes of the new regime. There is also the dependency risk which all groups run if they are drawn too closely into the policy process [*Olsen, 1983*]. Now that sustainable development has been written into the final regulations, the environmentalists risk being drawn into schemes which do not pass a central test which they have themselves long advocated. How will the environmentalists respond if the intellectual case for sustainable development is rejected at the local level, or if local political pressures are too great? The difficult challenge for the environmentalists is to develop sustainability indicators to show how, where, and if this is being achieved. At the same time, they need to maintain their long-standing policy focus, rather than becoming totally absorbed by the politics of implementation.

What then are the prospects for the apparently major changes in policy co-ordination procedures being translated into successful implementation? Two factors are likely to be crucial in determining policy outcomes. First, if issue salience can be maintained, it seems likely that the responses which we have described will continue. The issue will be difficult to ignore in practice. Secondly, the real test will be the extent to which policy implementors have really accepted the intellectual argument. If they have, then 'environmentalism' will become internalised and embedded in decision-making process. In such a situation, implementation failure is less likely. Thus, one of the most likely conditions for implementation success is when all policy actors subscribe to the basic goals of the programme. As Lundquist [*1972, 33*] suggests, those involved in implementation, if they do 'obey', do so for three main reasons.

1. The implementor obeys for fear of penalty or wish for reward.
2. The implementor obeys because he believes the decision to be rational, or because it agrees with his evaluation.
3. The implementor obeys because he appreciates the decision-maker personally, or because he always obeys the steering communications from certain organization roles, irrespective of who occupies the roles.

If the intellectual case has been won, then Lundquist's second factor will apply. Even if the intellectual case is still to be won, however, the 're-steering' that has been taken place since 1988–89 has begun to change the implementation situation from one in which resources were being provided but with a rather general mandate, to one in which resources (approximately £40 billion!) will be provided but with a much more specific mandate. As Montjoy and O'Toole argued, drawing upon US experience, '... the surest way to avoid intra-organisational problems is to establish a specific mandate and provide sufficient resources' [*Montjoy and O'Toole, 1979: 47*]. Finally, as problems of policy co-ordination are being pushed more to the regional and local level, there are some grounds for optimism since it may be that the practical environmental consequences of development decisions are apparent when those decisions are taken closest to the actual problem, rather than in Brussels. Whether this proves to be the case will depend on much broader questions relating to the development of intermediate governance structures in the EU and the meaning attached to the principle of subsidiarity. The policies discussed here will almost certainly be influenced by the uneven development of federalist structures and regional autonomy within the Union, as well as by the impact of subsidiarity upon EU environmental policy. These developments, depending upon what form they take may have either a positive or negative effect upon the coordination of EU environmental and regional policy.

ACKNOWLEDGEMENTS

This paper is part of a research project on lobbying in the EU, funded by the ESRC. The authors wish to thank all EU Commission officials and group representatives who agreed to be interviewed by us in connection with this research. They would also like to acknowledge the helpful comments of Laura Cram, Andrew McLaughlin, the co-editors and an anonymous referee.

NOTES

1. For details of the development of EU regional policy and the 1988 reform of the Structural Funds see: Bohan, N., 1992, 'Cohesion and the Structural Funds' in P. Ludlow, J. Mortensen and J. Pelkmans (eds.), *The Review of Community Affairs 1991* (Brussels: Brassey's), pp.266–275; Doutriaux, Y., 1991, *La Politique Régionale de la CEE* (Paris: Presses Universitaires de France).
2. Objective 1: lagging regions where GNP per capita is no more than 75 per cent EU average; Objective 2: declining industrial regions; Objective 3: assistance for long-term unemployment; Objective 4: integration of young people into the labour force; Objective 5: adjustment of agricultural structures and rural development.

REFERENCES

Commission of the EC, 1993a, 'Integrating the Environment into other Policy Areas within the Commission', *Press Release*, IP(93) 427, Brussels, 2 June.
Commission of EC, 1993b, 'The Structural Funds for 1994–1999', *Background Report*,

ISEC/B16/93, 28 May.
Commission of the EC, 1991, 'The New Structural Policies of the European Community', *European File,* June–July.
Commission of the EC, 1990a, *Environmental Policy in the European Community,* Luxembourg.
Commission of the EC, 1990, 'ENVIREG', *Info Technique,* Brussels.
Commission of the EC, 1989, *1992: The Environmental Dimension, Task Force Report on the Environment and the Internal Market.*
Court of Auditors, 1992, 'Special Report No. 3/92 Concerning the Environment Together with the Commission's Replies', *Official Journal,* 92/C245/01, Vol.35, 23 September.
Doutriaux, Y., 1991, *La Politique Réionale de la CEE* (Paris: Presses Universitaires de France).
Downs, A., 1967, *Inside Bureaucracy* (Boston: Little, Brown & Co).
Downs, A., 1973, 'The Political Economy of Improving Our Environment', in J.S. Baine (ed.), *Environmental Decay: Economic Courses and Remedies* (Boston: Little, Brown & Company), pp.59–81.
Gregory, R., 1971, *The Price of Amenity* (London: MacMillan).
Judge, D., 1992, 'A Green Dimension for the European Community', *Environmental Politics,* Vol.1, pp.1–13.
Ludlow, P., J. Mortensen and J. Pelkmans (eds.), *The Review of Community Affairs 1991* (Brussels: Brassey's).
Lundquist, L., 1972, 'The Control Process: Steering and Review in Large Organisations', *Scandinavian Political Studies,* Vol.7, pp.29–34.
Marks, G., 1992, 'Structural Policy in the European Community', in A.M. Sbragia (ed.), *Euro-Politics Institutions and Policymaking in the 'New' European Community* (Washington D.C.: The Brookings Institution), pp.191–224.
Majone, G., 1989, *Evidence, Argument and Persuasion in the Policy Process* (New Haven: Yale University Press).
Mazey, S. and J. Richardson, 1992, 'British Pressure Groups in the European Community: the Challenge of Brussels', *Parliamentary Affairs,* Vol.45, pp.92–127.
Mazey, S. and J. Richardson, 1993, 'Introduction' in S. Mazey and J. Richardson (eds.), *Lobbying in the European Community* (Oxford: Oxford University Press).
Montjoy, R.S. and L.J. O'Toole, 'Toward a Theory of Policy Implementation: An Organisational Perspective', *Public Administration Review,* Vol.39, No.5, pp.465–76.
Olsen, J.P, 1983, *Organised Democracy: Political Institutions in a Welfare State – The Case of Norway* (Oslo: Universitetsforlaget).
Peters, G., 1992, 'Bureaucratic Politics and the Institutions of the European Community', in A.M. Sbragia (ed.), *Euro-Politics Institutions and Policymaking in the 'New' European Community* (Washington D.C.: The Brookings Institution), pp.75–122.
Schmitter, P.C., 1990, 'Modes of Sectoral Governance: a Typology', Unpublished paper, Stanford University.
WWF, 1989, *Reform of the Structural Funds: An Environmental Briefing* (WWF: Godalming, Surrey and London Institute for European Environmental Policy).
WWF/IEEP, 1990, *The EC Structural Funds – Environmental Briefing, 2* (WWF UK: Godalming, Surrey).
WWF, 1992a, *Statement on Behalf of European Environmental Groups: Reform of the EC Structural Fund Regulations* (Brussels: WWF European Office).
WWF, 1992b, *Briefing for the Danish Presidency* (Denmark: WWF).
Weale A., and A. Williams, 1992, 'Between Economy and Ecology? The Single Market and the Integration of Environmental Policy', *Environmental Politics,* Vol.1, pp.45–64.

Policy Networks on the Periphery: EU Environmental Policy and Scotland

ELIZABETH BOMBERG

I. INTRODUCTION

This chapter employs a 'policy networks' model to analyze the role of peripheral regions in European Union (EU) environmental policy-making. It argues that a policy networks approach provides clues for understanding both the EU's general environmental policy-making process, as well as the particular role of the peripheral regions in this process. The chapter begins by outlining the basic components and variations of functional 'policy networks'. Section II suggests that EU environmental policy may be viewed as being made within a relatively loose and permeable 'issue network'. Section III analyses the Scottish case to illustrate the main components and dynamics of 'territorial communities' located in the periphery. Section IV examines the interaction of these two networks, arguing that whilst the Scottish territorial community is sometimes able to represent its interests in EU policy-making, it is ultimately constrained by its 'embeddedness' in the wider framework of UK governance. Section V considers how the constellation of power within and amongst policy networks can change.

Policy Networks

Policy networks are primarily a means for describing relationships between private and public actors. The approach is an attempt to move beyond pluralist and corporatist models of government/private interest intermediation. Like pluralism and corporatism, the policy network approach attempts to explain the intermediation of public and private interests. But analysts employing a policy network approach argue that both pluralism and corporatism offer only a very general, unspecific model of these relationships. Jordan and Schubert point out that states are not uniform in their capacities in all policy areas [*1992: 10*]. In 'pluralist' systems, some fields of sectoral corporatism will exist, while in corporatist systems there are policy areas that exhibit a pluralist pattern of interest intermediation. In other words, the relationships between private and public interests vary considerably across policy areas. The policy networks approach emphasizes the need to disaggregate policy analysis because of this variation [*Marsh and Rhodes, 1992: 4*].

Policy networks have been defined as a cluster of 'organizations connected to each other by resource dependencies; and distinguished from other clusters by breaks in the structure of resource dependencies' [*Benson, 1982: 148*]. The term, 'resource dependencies' refers to mutual need for resources such as information, expertise, access and legitimacy; 'breaks' refers to the fact that specific resources are compartmentalized and dedicated to a specific policy area or set of goals.

A policy network thus includes a relatively stable set of public and private actors. The dependencies (or linkages) between actors serve as channels for communication and exchange of information, expertise, trust and other policy resources. The boundary of a policy network is not determined by institutions or formal distribution of power but by a mutual recognition of resource dependencies [*Kenis and Schneider, 1989: 14*]. In other words, policy networks emphasize the informal relationships surrounding policy-making. Typically, bureaucrats from different levels of government, interest groups and committees of experts are intimately involved in setting policy agendas, defining policy problems, and setting out an acceptable range of options. The policy networks approach assumes that these informal factors often better account for policy outputs than do party stances, political leadership or parliamentary influence.

This chapter adopts the concept as refined by Rhodes [*1988*; *Marsh and Rhodes, 1992*]. In the 'Rhodes model' the following factors are emphasized:

1) the relative stability of network membership;
2) resource dependencies;
3) the relative insularity and autonomy of a network from wider influences.

In most cases, policy networks form around specific policy sectors (agriculture, environment, etc.) or policy functions (regulation, implementation, etc.). Working with the three variables listed above we can construct a continuum of functional networks (see Figure 1). At the 'tight' end of the continuum are policy communities. These are networks characterized by stable relationships, restricted membership, strong structural dependencies, and insulation from other networks, parliament, and the general public. Policy communities act to 'depoliticize' a policy arena by excluding groups which are likely to disagree with the established policy agenda [*Smith, 1991: 235–6*]. The most cited example of a policy community at work is the network surrounding British or EU agricultural policy-making [*Smith, 1991*; *Peterson, 1989*].

FIGURE 1
CONCEPTUALIZING POLICY NETWORKS

Policy Communities	*Issue Networks*
←	→
Stable membership	Fluid Membership
Highly insular	Highly permeable
Strong dependencies	Weak dependencies

At the other end of the continuum are issue networks, characterized by fluid membership, highly permeable boundaries, weak dependencies, and the inclusion of wider, oppositional interests. Heclo [*1978*] views issue networks as incorporating a wide range of decision-making centres and a wide range of actors which move in and out of policy arenas and have different interests. Compared with other networks, an issue network is a relatively *ad hoc* policy-making structure in which not only a large, but to a certain extent unpredictable number of conflicting interests participate. A more detailed illustration of an issue network is provided below.

II. EU ENVIRONMENTAL POLICY AS AN ISSUE NETWORK

The notion of policy networks can be used to explain policy-making in national settings as well as in the EU. This transnational application is relatively recent. Peterson [*1989*] employs the model to analyze aspects of the EU's agricultural policy. Others use the approach loosely to help explain the strategies and effects of lobbying in the EU [*Mazey and Richardson, 1992*].

The approach is useful at the EU level because it recognizes the complexity of the EU policy-making system. The EU incorporates a wider set of interests than does any national system of policy-making. Moreover, compared with national settings, power within the EU is highly situation-specific, with few clear lines of authority or standard consultative patterns. The process of policy formation in the EU is multi-layered and complex. The policy network approach can be used to highlight what is different about EU policy-making in discrete sectors of policy. It draws attention to the varying strength of different actors in different policy sectors. Its emphasis on resource dependencies helps identify which actors' interests are decisive and why.

The approach is a particularly helpful tool in understanding the relatively new system of EU environmental policy-making. Unlike more entrenched policy sectors such as those surrounding the regulation of the chemicals industry or agricultural policy, patterns of EU policy-making in the environmental arena are still quite new and fluid. Environmental policy is characterized by rapid change and fierce struggles between competing institutions and actors. It is best understood as occurring within an issue network populated by a wide range of actors. A brief examination of EU environmental policy-making underlines this point.

The first characteristic of a network is its permeability, the extent to which external pressures impinge on the network's policy-making process. Growing public concern for the environment has 'politicized' environmental issues and rendered them susceptible to public pressures, debate and scrutiny. Moreover, environmental issues have become increasingly intertwined with other sectoral issues, especially those surrounding agriculture, regional funding, transport and internal market policies. Pressures from these other sectors influence the scope and nature of environmental policy and blur the boundaries of the environmental issue network.

Indeed, most environmental issues cross over several policy areas and EU Directorates-General (DGs). Extensive discussions must take place within the Commission itself to determine which policies should be allocated among which Directorates. In short, the network's permeability means that 'there is not only conflict regarding outcomes but about the definition of the problem' [*Jordan and Schubert, 1992: 13*].

Because of the wide and diverse scope of issues covered in the environmental network, membership tends to be unstable or fluid. The EU's environmental issue network can include actors from DG XI (the Directorate responsible for Environment, Nuclear Safety and Civil Protection), members of other DGs, national civil servants and representatives, scientific experts, Members of the European Parliament (MEPs), environmental and business interest groups. In this complex and crowded policy-making *milieu* there is a changing cast of participants over time and according to specific issues. Whilst a small group of officials in DG XI provide some form of membership stability, they are but one of many actors involved in the policy-making process.

Given the restricted size of its permanent staff, DG XI must rely on a wide array of participants from outside its department for technical and political advice. In particular, it depends on experts and officials on secondment from other EU institutions, from member-state departments, and from private organizations and foundations. One official in DG XI claimed that 'we have more officials on loan than any other DG'.[1]

The strength or weakness of resource dependencies represents a third defining component of policy networks. Instead of a tight relationship underpinned by strong dependency relationships, the environmental network includes more informal, less standardized sharing of resources among a wide variety of members. Network members located in DG XI rely on a broad range of sources (environmental groups, business interests, scientific experts, the administrative agencies in member-states, etc.) for information, expertise and legitimacy. They thus are not dependent on a select few sources. Conversely, interest groups concerned with environmental issues have several alternative access points located within the multiple EU institutions, as well as those on the national level. In short, dependencies remain weak because they are variable and diffuse.

An issue network is characterized also by the inclusion of a wide array of members and interests. For instance, the EU environmental issue network usually reflects the competing interests expressed by representatives of 12 member-states [see *Bomberg and Peterson, 1993: 148*]. Secondly, DG XI is generally considered to be more open to lobbyists than any other DG. In particular, the relatively open and diffuse nature of an issue network allows environmental non-governmental organizations (NGOs) to lobby and exert influence. These NGOs are sometimes successful in pushing issues on to the environmental agenda. An advisor to the Environment Commissioner observed that, especially in areas related to nuclear energy, waste transport and water management, lobbying from environmental groups 'clearly has some impact on the formulation of EC legislation'.[2]

Moreover, whereas policy communities tend to exclude parliamentary influence and scrutiny, the EU's environmental issue network is distinctively open to influence from the European Parliament (EP). Indeed, the Commission welcomes the Parliament's view precisely because the EP is democratically elected and thus can legitimate EU environmental policy. An official in Britain's Department of the Environment unit responsible for European affairs confided that 'the Parliament carries a lot of weight with the Commission because the Commission knows that even if [the EP's] powers are primarily advisory, it is democratically elected and the Commission is not'.[3]

The EP has successfully taken advantage of the greater environmental awareness among its electorates. Pushed by a small but vocal Green group of MEPs and an active Environment Committee, the Parliament has promoted its own 'environmental agenda' by providing the Commission with well researched inputs into legislative proposals on a variety of environmental issues, such as drinking water quality and auto emissions [*Bomberg, 1992*].

The comparatively open character of the EU's environmental issue network allows the EP and NGOs to play a role in agenda-setting, but this openness should not be exaggerated. First, the EP is not yet a legislative body; its power relative to other EU bodies remains weak. Moreover, patterns of EU environmental policy-making resemble national patterns in important ways; certain groups and interests clearly have more access than others. As Peterson argues '[t]he vast resources and specialized expertise needed to wield influence at the EC level may mean that the power of groups which already control the policy agenda in national setting often ends up being reinforced and extended'[*forthcoming: 37*].

Compared with other policy sectors, however, EU environmental policies are subject to bargaining among a wide array of actors over the definition of the problem as well as the desired outcome. This process usually includes, not only competing DGs and Commissioners and experts from various backgrounds, but also various EP committees, and representatives of national governments and interest groups.

III. THE TERRITORIAL DIMENSION: THE EXAMPLE OF SCOTLAND

The most common division among policy networks is along policy or functional lines; different policy networks form around different policy sectors. But there also may be a territorial division. For instance, Rhodes [*1988*] regards the territorial ministries in Scotland, Wales and Northern Ireland as distinctive species of sub-central government. These territorial ministries can serve as a centre around which new types of policy networks or 'territorial communities' appear. The Scottish case helps illustrate why these territorial communities are described as 'integrated policy communities'. Like policy communities, territorial communities tend to be characterized by stable membership, strong resource dependencies, and insulation from oppositional groups, parliament and the general public. They can thus be located towards the left end of the continuum shown in Figure 1. But they differ from policy communities in their multi-functional remit, their informality and their geographic peripherality.

Midwinter, *et al* agree that certain key characteristics of a territorial community, such as stable membership, insularity, strong resource dependencies, 'catch the Scottish style very well' [*1991: 201*]. To begin with, membership within the Scottish territorial community is relatively stable. Providing stability at the centre of the community is the Secretary of State for Scotland and the Scottish Office. As a member of the British Cabinet, the Scottish Secretary of State can put before the Cabinet the Scottish viewpoint on major matters of policy. However, the Cabinet itself spends little time on Scottish policy matters. Full Cabinet meetings do not usually discuss detailed policy matters of relevance to Scotland. Such discussions are left to the relevant Cabinet committees of which the Scottish Secretary may or may not be a member. Moreover, Keating and Midwinter [*1982: 3*] note that the Secretary of State's membership in Cabinet also carries with it the obligation of collective responsibility for the actions of government. As a result, the Scottish view is sometimes compromised in order to reach collective decision. Even with these restrictions, the Secretary of State's influence is substantial and he remains a dominant actor in the territorial community. He links the territorial community to the highest level of executive power.

Other members of the territorial community include Scottish Office officials and UK civil servants in Whitehall, scientific experts, and interest groups. Local authorities can play an important role in the implementation of policies, but their role in policy formulation is still limited [*Midwinter, et al., 1991: 148ff*]. Although membership of this community varies somewhat according to the issue, the primary movement of members tends to occur *within* the community rather than 'in and out'. For instance, civil servants within the Scottish Office often switch departments and portfolios. But membership of the network itself remains coherent. As one Scottish Office official put it, 'We often move rooms but we seldom move house'.[4]

The feature which most distinguishes a territorial from a functional network is the former's responsibility over a wide range of different functions or policy sectors. Instead of administering one function the Scottish Office is responsible for a range of policies for a single part of the country. The Scottish Office has administrative responsibility for most functions, excluding treasury, foreign and defence affairs. It consists of several departments including Agriculture and Fisheries, Industry, Environment, Education, Home and Health, and Development. A British Forestry Commission is also based in Scotland. All departments answer to the Secretary of State for Scotland. Because territorial communities penetrate a range of policy arenas, policy formulation involves the constant interaction among different policy issues and actors within the network. Put another way, the multi-functional remit of territorial communities requires that the policy process also be multi-functional. For example, funds for conservation projects are not exclusively controlled by the Environment Department but rather are routinely vetted by civil servants in other departments such as Industry or Agriculture.

A third distinguishing characteristic of a territorial network is its atmosphere of informality and close communication. The Scottish community's comparatively small size allows for an intimacy not possible in a large scale network. Moreover, the fact that they are located on the periphery can instil a sort of 'village voice mentality' among territorial community members. This informality is also due to the community's transectoral character which encourages continual contact and subsequent familiarity. In particular, informal and continual communication is necessary to reach a coherent 'Scottish position' on national and supranational issues. As one Scottish Office official noted, 'There has to be a lot more communication. Because you all work under the same minister of Cabinet, you have to resolve things which in England do not get resolved in anything like the same way'.[5]

The Scottish case reveals a fourth characteristic of territorial communities: strong resource dependencies. The territorial community's smaller size and peripherality means members have a limited resource base on which to depend. The resources of expertise, co-operation, access and information tend to emerge from and remain within the territorial network. Participants in this resource exchange can include colleagues in Whitehall but these personal contacts are usually kept to a minimum [*Keating and Midwinter, 1982: 3*]. Similarly, there is a tendency to rely on a pool of scientific experts within the Scottish network rather than 'bring up someone from down South'.[6] Finally, members of the community are generally forced to depend on one another to represent Scottish interests beyond Scottish borders. The most effective way to carry weight in the Cabinet or the EU is to have a common, unified position.

A further characteristic of territorial communities is their insularity. More specifically, territorial communities are known for their exclusion of parliamentary participation, opposition groups and the general public. Midwinter, *et al.* argue that '[g]iven the British tradition of secrecy and executive dominance in government . . . much of the process of territorial bargaining and accommodation is conducted away from the public gaze' [*1991: 202*].

Certain interest groups do have access to the territorial community. A Scottish Office official noted a regular contact between the Scottish Office and business interest group representatives from the Confederation of British Industry (Scotland).[7] But representatives of oppositional environmental groups such as Friends of the Earth (FoE) Scotland do not enjoy the same access. The Director of FoE Scotland claimed that environmental groups 'in London and Brussels have far less trouble than we do permeating the policy making circles. The circle of power here is impenetrable . . .'.[8] This may be an overstatement, but Scottish environmental groups organized also at the national or European level (such as the World Wide Fund for Nature or the Royal Society for the Protection of Birds) tend to have greater success when lobbying the more permeable networks in Whitehall and Brussels.

Parliamentary participation is also limited in the Scottish territorial community. First, the vast majority of Scottish Members of Parliament (MPs) are from parties opposing the Conservative government and thus may lack extensive political or personal connections with the Scottish Office or with a Secretary of State who is an elected Conservative MP. More generally, Keating and Midwinter observe that although MPs may suggest policy initiatives, the Scottish Office, and particularly the Secretary of State 'plays the major role in determining which of these issues will reach the political agenda and what their priority will be' [1982: 8].

A final characteristic of a territorial community might be termed 'territorial embeddedness'. As subnational units, territorial communities remain part of a larger institutional and political framework. The overarching structure of any political system – particularly how powers are distributed between different institutions – will set the parameters within which territorial communities can operate.

The Scottish territorial community's operations reflect power relationships between institutions in the wider UK political system. There is no Scottish assembly to represent Scottish interests: power in the UK is centred in the Cabinet where only one minister is responsible for Scottish affairs. As part of the UK government, the Scottish Office cannot conclude agreements with international bodies such as the EU 'other than in the capacity as a branch of the UK central government' [Mazey and Mitchell, 1993: 109]. In short, the territorial community remains subsumed within wider networks operating on a higher level. This 'embeddedness' can greatly limit the community's ability to manoeuvre.

In sum, in contrast with the loose network characterizing EU environmental policy, the Scottish territorial community represents a multifunctional, integrated, stable, informal network with continuity of membership, a high degree of insularity and significant constraints resulting from its embeddedness in the wider political system. This application of the policy network approach helps us understand which powers dominate policy-making in different environments, and why. But to understand the role played by peripheral communities in EU policy-making, the following section examines the interaction between the Scottish territorial community and the EU's environmental issue network.

IV. INTERACTING NETWORKS

Territorial communities enjoy distinct advantages accruing from their peripheral status. First, because of their size and peripheral nature, these communities develop a level of co-ordination and solidarity not present at the UK or EU level of policy-making. The achievement of coherency is also facilitated by the community's informality and insularity. For instance, the need for departments within the Scottish Office to work together on UK policy formulation is even greater in cases related to EU policies. Most Scottish input on EU policies are routed through London to UKREP, a collection of UK officials in Brussels who represent the Whitehall departments most affected by EU policies and who work under the direction of the UK 'ambassador' to the EU. There is a general recognition that for the 'Scottish interest' to be represented, the network must present a coherent front. This imperative serves to rally members around a common purpose: 'Our stance must be united and coherent enough so that the UK representatives can take it forward to UKREP and Brussels'.[9]

Secondly, a territorial community can take advantage of its small and distinctive character. For instance, its small size makes personal relationships easier to cultivate and maintain [*Wilks and Wright, 1987: 274–313*]. These include relationships with those involved in the EU environmental policy-making network, i.e. those working in the Commission or European Parliament. In the Scottish case, much use is made of personal relations amongst Scots working within the EU. One Scot working for the Environment Commissioner underlined the importance of these contacts: 'The fact is that within the decision-making procedure there are a large number of Scots. There's me, there's Scots in DG XI and in UKREP. And we have made it our business to ensure that the Scottish point of view is heard. And when we consider that the UK government is not taking sufficient account of Scottish interests, we tell them that.'[10] Impromptu gatherings are sometimes held in Brussels where Scots can meet on an informal basis. There even exists a handbook of 'Jock Tamson's Bairns' which lists Scots working in EU institutions and is designed to facilitate the use of personalized networks.

These personal relationships among Scots can provide an indirect representation of territorial interests in Brussels. But territorial communities can also work more directly by 'inject[ing] in central governments a territorial component which involves . . . substantive policy variations' [*Rhodes, 1988: 285*]. At the level of EU policy formulation, the Scottish territorial community can influence policy through its direct participation in EU policy networks. First, UKREP includes at least one official from the Scottish Office at all times. Although centred in Brussels this official refers back to Edinburgh for guidance. On matters of concern to the network, there may be an additional Scottish Office official attached to the entourage of the UK minister in Brussels [*Midwinter, et al., 1991: 87*].

If the members of the network deem an issue particularly relevant to Scotland (i.e. policies on hill farming or afforestation) the network can take on a larger role. A recent example occurred in the formulation of the

1992 Habitats Directive (Council Directive 92/43/EEC). This directive establishes 'sites of Community importance', which are designated as those which contribute significantly to the maintenance of biological diversity or a 'natural habitat type'. The Scottish community developed a keen interest in this directive because a large number of potential sites are located in Scotland. In this case an entire team of Scottish Office administrators and experts were included in the COREPER (Committee of Permanent Representatives) working group. These working groups can have a decisive influence on the details of EU policy decisions.

In some instances, the Secretary of State can represent the territorial community by personally taking part in Council of Ministers meetings. For example, following the January 1993 Shetland oil spill, the Secretary of State for Scotland, Ian Lang, attended a joint Council meeting of Environment and Transport ministers in Brussels. The meeting included a discussion of the oil spill's effect on Scotland's natural habitats, fisheries and tourism, as well as the possibility of further EU funding to help meet clean-up costs in Shetland. More general policies were also discussed such as maritime navigation and safety. In sum, there are important instances in which members of the territorial community are involved in determining policy priorities and in shaping policy responses.

It is easy, however, to exaggerate the case of Scottish influence. Overall the Scottish Office's role in EU policy-making has been largely reactive, responding to initiatives from Whitehall or Brussels rather than taking the lead itself. Secondly, the personal links among Scots are erratic and unpredictable; they can not substitute for the institutionalized representation on the Council of Ministers. Moreover, the importance of 'personal connections' on policy formulation is probably overstated by network participants.[11]

Many of the advantages listed above (size, intimacy, consensual imperative) can be disadvantages in different situations. For instance the network's comparatively small size can also limit its impact. In presenting a Scottish network proposal to civil servants in Whitehall, Scottish officials confront severe problems of scale. As Keating and Midwinter note, 'One Scottish Office official will face several officials from a Whitehall department covering the same area; and both junior ministers and officials will find that their Whitehall counterparts with similar breadths of responsibility will be considerably senior. This inevitably tends to reduce the Scottish impact on policy' [*Keating and Midwinter, 1982: 15*].

Whilst peripherality and 'distinctiveness' may have its advantages, administrators in the Scottish territorial community regularly express indignation at being treated as mere 'peripheral interests' by their British or European colleagues. For instance, several conservation scientists and officials from the Scottish network became closely involved in the formulation of the 1992 Habitats Directive. But there was a general disgruntlement among Scottish Office representatives that they were not treated as 'equal partners' to their Whitehall counterparts in these discussions.

The most serious impediment to community influence, however, concerns the territorial community's position in the wider system of UK government. Constitutional and political imperatives render direct Scottish participation difficult. When asked about direct contact between the Scottish territorial community and the Commission, one advisor to the Environment Commissioner noted: 'I think DG XI and those here on the political level would quite like to speak directly to the Scottish Office. But it is the United Kingdom which decides which way it's going to go, how the lobbying is going to be done'.[12]

In other words, the Scottish Office is only a minor actor in wider UK policy networks. Although central in the Scottish network, the Scottish Office remains a department of central government staffed by civil servants and headed by UK Cabinet minister [*Keating and Midwinter, 1982: 2*]. The need to agree on a UK-wide position before EU sessions, and to abide by the negotiated outcomes, leaves only limited scope for separate Scottish policies. This situation will not be altered without constitutional change in the UK. Such change is not a priority of the current government whose position on the issue is that 'some things in politics are non-negotiable'.[13]

If the territorial community's influence is limited, the influence of those outside the community is doubly marginalized. The above analysis has revealed that interests must first be filtered through the territorial network and moulded into a coherent position before they can hope to influence UKREP and Brussels. Those deprived of access to the community in the first place are at an obvious disadvantage. Put simply, the territorial community itself remains an elite. It is open to other governmental and department interests, open to certain interest groups and a variety of EU actors, but relatively closed to opposition interests such as environmental groups or opposition parties. Rhodes [*1988: 286*] concludes that territorial communities act to sustain existing policies and patterns of distribution. They are not champions of oppositional interests, nor are they agents for change.

This marginalization is furthered by the close personal relationships which facilitate the interaction between Scottish and EU networks but rarely extend to members of oppositional groups. Nor are peripheral local authorities normally involved in the formulation of policies which ultimately affect them. For example, the Highlands and Islands local authorities will be significantly affected by the EU's Habitats Directive, as a disproportionate number of potential sites are located in this region.

But according to one Scottish Office administrator involved in the Directive, 'there was really little local authority interest or capacity to involve themselves in negotiation of the [Habitats] Directive . . . and little need for central government to be consulting them'.[14]

The point is, networks do interact. Certain 'Scottish interests' may find their way from the Scottish territorial community to the EU environmental issue network. But on the whole, these inputs tend to be random, 'one-offs' rather than a reflection of systematic consultation or shared resources. Moreover, the interests of oppositional environmental groups and local authorities are often 'squeezed out' in the process.

Consequently, the formulation of environmental policies for the periphery may exclude peripheral actors. The more decisive interaction seems to be that between British and EU policy networks. For example, the UK has agreed to drinking water directives which require the regular testing of water purity. The directive was seen as an unfair and unnecessary administrative burden on regional authorities in the Scottish Highlands where, according to one Scot working in the Commission, 'the only thing contaminating water is an occasional dead sheep'.[15]

V. NETWORK CHANGE?

Policy networks – whatever their type – can change [see *Marsh and Rhodes, 1992: 257ff*]. The impetus for change is unlikely to come from within the Scottish territorial network itself. Scott *et al.* *[1994]* argue that rather than being an agent for change, the Scottish Office has 'increasingly become an agent for ensuring that centrally determined policy goals are not altered or challenged at the local level'.

Consequently, the primary push for change is more likely to come from 'the bottom up'. This would involve those 'on the margin' exploiting their peripheral status to their own advantage. In particular, oppositional groups and regional and local authorities might by-pass both the Scottish community and Whitehall by asserting their presence directly in Brussels. To a certain extent this is already occurring. For example, regional councils such as Strathclyde and Highlands and Islands have established direct representation in Brussels. They have proven adept at securing regional funds from the Commission, even though all applications must first be vetted in Whitehall. Similarly, in 1991 the organization 'Scotland Europa' was established in Brussels to enable Scottish local authorities and business to lobby the EU more effectively. Oppositional environmental groups in Scotland are beginning to take seriously the advantages of organizing on the European level.[16] In short, '[t]he territorial opposition is beginning to see Brussels as more accessible and receptive to their demands than is Whitehall' [*Keating and Jones, 1991: 322*].

Change might also come from 'above' as the Commission seeks to increase communication and contact with subnational groups and regional bodies on the periphery. In 1988 the Commission established a consultative council of regional and local authorities to encourage these links. Scottish local government is part of this consultative network. The Maastricht Treaty on European Union provides for a Committee of the Regions and Local Authorities, which will consists of 189 representatives from local and regional governments, including 24 from the UK. The Committee is designed to increase substantially the consultative role of regional and local interests in the EU decision-making process.

Moreover, the Commission has encouraged direct links with interest groups within the UK. It has set up a Commission office in Scotland's capital city of Edinburgh. According to Midwinter *et al.*, the office 'reflects a recognition by the Commission of the special position of the peripheral nations of the UK' and is designed to help interested parties through the maze of EU regulations and grants [*1991: 88*].

The effect of such dialogue also raises the possibility of a new domestic relationship between the Scottish and wider British functional networks. For instance, Anderson [*1990: 422*] argues that EU actions which strengthen the regions enable subnational groups to redefine the terms of their exchange relationships on the national level. Anderson suggests that by altering the distribution of resources among domestic networks, the EU may become a source of domestic empowerment for actors within territorial communities.

These moves by peripheral actors and the Commission suggest the long-term possibility of network boundaries being redrawn, or new networks being constructed. But clearly, several factors may stymie prospects for increased collaboration between the Commission and peripheral or oppositional interests. Midwinter *et al.* [*1991: 88*] argue that because of the absence of an elected government at the Scottish level, Scottish representatives lack political stature and influence in Brussels. More importantly, national governments will jealously protect their positions as intermediaries between subnational interest and Brussels [see *Anderson, 1990: 423*]. The establishment of a coherent and effective transnational representation of environmental interests is still a long way off.

CONCLUSION

This chapter has presented the policy networks approach as a useful way of explaining the relationship between peripheral regions and EU environmental policy-making. The application of this model has revealed its strengths as well as weaknesses. The primary purpose of network analysis is to explain the way in which meso-level decision-making affects policy outcomes. It is of limited use in explaining 'macro' developments which may impinge on the role of peripheral regions in EU policy-making. For instance, this chapter has highlighted the possible changes in role of regions resulting from the Maastricht Treaty's creation of a Committee of Regions. The link between the everyday decisions and the larger, 'macro' developments needs to be examined.

The importance of 'macro' or structural constraints is particularly important for territorial communities. Research on territorial networks has tended to de-emphasize or neglect the extent to which territorial communities are embedded within other networks on higher policy levels. Whilst the Scottish case illustrates the opportunities enjoyed by territorial communities, it also highlights significant structural or constitutional constraints. These 'macro' constraints clearly limit the policy-making influence of peripheral actors. Further research is also needed on the interaction among different networks, especially the interaction between the functional areas of environmental and regional policy. This chapter has illustrated how, for the peripheral regions in particular, this interaction can have significant consequences for environmental protection and development. Research is needed which makes explicit how decisions are co-ordinated across different policy sectors as well as among levels of government.

Overall, however, the model of policy networks can be extremely useful in explaining the role of peripheral regions in EU policy-making. The policy network approach helps us order evidence in novel ways and it provides clues to anticipate policy outcomes. In particular, it helps explain how decisions are made in different settings, which actors are decisive and why.

Beyond that, the approach encourages the exploration of the interaction among networks. This study has examined the intersection of territorial and functional networks. It was revealed that this interaction provided limited advantages for peripheral territorial communities, and that the interests of local and oppositional groups were marginalized in the process. Barring significant network change, we can thus anticipate that EU environmental policy formulated for the periphery will not express peripheral interests. Rather, policy outcomes will continue to reflect the bargaining between multiple actors at the EU level and at the level of central governments. In the area of EU environmental policy, the impact of territorial communities will remain peripheral.

ACKNOWLEDGEMENT

The author would like to thank Dave Marsh, John Peterson, Rod Rhodes, the editors and an anonymous reviewer for their helpful comments.

NOTES

1. Interview with official in DG XI, 27 January 1993, Brussels.
2. Interview with member of Environment Commissioner's *cabinet*, 28 January 1993, Brussels.
3. Interview with official in the UK Department of Environment, 11 September 1991, London.
4. Interview with Scottish Office official, 22 January 1993, Edinburgh.
5. Interview with Scottish Office official, 5 February 1993, Edinburgh.
6. Interview with Scottish Office official, 5 February 1993, Edinburgh.
7. Interview with Scottish Office official, 5 February 1993, Edinburgh.
8. Interview with Director of Friends of the Earth Scotland, 3 February 1993, Edinburgh.
9. Interview with Scottish Office official, 5 February 1993, Edinburgh.
10. Interview with member of Environment Commissioner's *cabinet*, 28 January 1993, Brussels.
11. Indeed, one Scottish MEP insisted that because of his 'central' position in the EU's environmental policy network 'you can hardly say that Scotland is politically peripheral' (Interview with Scottish MEP, 26 February 1993, Scotland).
12. Interview with member of Environment Commissioner's *cabinet*, 28 January 1993, Brussels.
13. Prime Minister John Major quoted in *The Scotsman*, 30 January 1993.
14. Interview with Scottish Office official, 22 January 1993, Edinburgh.
15. Interview with member of Environment Commissioner's *cabinet*, 28 January 1993, Brussels.
16. Interview with Director of Friends of the Earth Scotland, 3 February 1993, Edinburgh.

REFERENCES

Anderson, J., 1990, 'Skeptical Reflections on a Europe of Regions: Britain, Germany and the ERDF', *Journal of Public Policy*, Vol.10, pp.417–47.

Bomberg, E. and J. Peterson, 1993, 'Prevention from Above? Preventive Policies and the European Community' in M. Mills, (ed.), *Health Prevention and British Politics* (Aldershot: Avesbury Press), pp.139–59.

Bomberg, E., 1992, 'The German Greens and the European Community: Dilemmas of a Movement Party', *Environmental Politics*, Vol.1, No.4, pp.160–85.

Benson, J.K., 1982, 'A Framework for Policy Analysis', in D. Rogers and D. Whitten (eds.), *Interorganizational Coordination* (Iowa: Iowa State University Press), pp.137–76.

Heclo, H., 1978, 'Issue Networks and the Executive Establishment', in A. King (ed.), *The New American Political System* (Washington, DC: American Enterprise Institute), pp.87–124.

Jordan, G. and K. Schubert, 1992, 'A Preliminary Ordering of Policy Network Labels', *European Journal of Political Research*, Vol.21, pp.7–27.

Keating, M. and B. Jones, 1991, 'Scotland and Wales: Peripheral Assertion and European Integration, *Parliamentary Affairs*, Vol.44, pp.311–24.

Keating, M. and A. Midwinter, 1982, 'The Scottish Office in the United Kingdom Policy Network', *Centre for the Study of Public Policy*, Studies in Public Policy Paper Number 96 (Glasgow: University of Strathclyde).

Kenis, P. and V. Schneider, 1989, 'Policy Networks as an Analytical Tool for Policy Analysis', Paper for a conference at the Max Planck Institute, Cologne, 4–5 December 1989.

Marsh, D. and R.A.W. Rhodes (eds.), 1992, *Policy Networks in British Government* (Oxford: Oxford University Press).

Mazey, S. and J. Richardson, 1992, 'British Pressure Groups in the European Community: The Challenge of Brussels', *Parliamentary Affairs*, Vol.45, pp.92–107.

Mazey, S. and J. Mitchell, 1993, 'Europe of the Regions: Territorial Interests and European Integration: The Scottish Experience', in S. Mazey and J. Richardson (eds.), *Lobbying in the European Community* (Oxford: Oxford University Press).

Midwinter, A., M. Keating and J. Mitchell, 1991, *Politics and Public Policy in Scotland* (Basingstoke: Macmillan).

Peterson, J., 1989, 'Hormones, Heifers and High Politics–Biotechnology and the Common Agricultural Policy', *Public Administration*, Vol.67, pp.455–71.

Peterson, J., (forthcoming), 'Governance in the European Union: Towards a Framework of Analysis', *York Working Paper in Politics*, University of York.

Rhodes, R.A.W., 1988, *Beyond Westminster and Whitehall* (London: Unwin Hyman).

Scott, A., J. Peterson and D. Millar, 1994, 'Subsidiarity: A "Europe of the Regions" vs the British Constitution?', *Journal of Common Market Studies*, Vol.32, No.1.

Smith, Martin, 1991, 'From Policy Community to Issue Network: Salmonella in Eggs and the New Politics of Food', *Public Administration*, Vol.69, pp.235–55.

Wilks, S. and M. Wright (eds.), 1987, *Comparative Government–Industry Relations: West Europe, the United States and Japan* (Oxford: Clarendon Press).

Administrative Capacity and the Implementation of EU Environmental Policy in Ireland

CARMEL COYLE

AN ISLAND ON THE PERIPHERY

The Irish people have consistently supported Ireland's[1] membership of the European Community (now European Union). In 1972 there was an 82 per cent vote in favour of Ireland joining the EC. Membership was seen as an opportunity to move away from economic dependence on the British market, especially for agricultural produce where Britain's cheap food policy depressed prices well below the continental European levels. Moreover, as a peripheral country with low economic development and a lower GDP (gross domestic product) per head than any other member state at the time, there was the prospect of eventual financial transfers from the Regional Fund which was being established at the time of Irish entry. Although support for the EC in later referenda on the Single European Act and the Maastricht Treaty fell to 70 per cent and 69 per cent respectively, there is still a broad consensus that EU membership remains the best option for Ireland.

As one of the less developed economies in the EU, Ireland was designated an 'Objective 1' region, thereby gaining maximum support from the Structural Funds which are designed to reduce regional disparities between member-states. Ireland has done remarkably well from EU funding, securing a higher proportion of funding per capita than either Greece or Portugal, whose GDPs are much lower. The *National Development Plan 1989–1993*, which the Irish government submitted to Brussels in March 1989 for Structural Fund support, identified several structural weaknesses likely to impede the capacity of the Irish economy to benefit fully from the opportunities arising from the completion of the Single Market:

a) low income and output levels – in 1988 per capita GDP was 62 per cent of the EC average;
b) a population structure resulting in rapid growth in labour supply and a high dependency ratio – about 66 per cent of the Irish population are in the under 15 or over 65 age-groups compared with an EU average of 50 per cent;

c) weak labour demand leading to an unemployment rate of around 18.5 per cent and persistent emigration, estimated at 2.5 per cent of the labour force in 1988;
d) constraints imposed by budgetary imbalances and public sector indebtedness – at the end of 1988 the national debt was the equivalent of 133 per cent of GNP (gross national product) compared to the EU average of 60 per cent;
e) high access and transport costs arising from the country's peripheral and island location;
f) a poorly developed infrastructure which hindered development and added to costs;
g) a heavy dependence on agriculture, which in 1988 accounted for around 11 per cent of GDP and 15 per cent of employment;
h) a weak industrial structure characterized by small scale traditional indigenous industries with weak business skills;
i) low levels of investment reflecting the weak structure and narrow base of indigenous industries [*Government of Ireland, 1989, Ch.1*].

The more positive economic tone in the *National Development Plan 1994–1999*, submitted in October 1993 for the current round of Structural Funding, reflected the stimulus to the economy of the financial transfers under the 1989–93 Community Support Framework – Ir£3.3 billion at current prices. The rate of economic expansion at five per cent was about three times the EU average and Ireland's GDP is projected to rise to 73 per cent of the EU average by the end of 1993. Inflation averaged three per cent over the period 1989–93 compared with an EU average of 4.75 per cent, while the balance of payments moved from a deficit of about 1.5 per cent of GNP to a surplus of six per cent of GNP in 1993. However, while overall levels of employment have been maintained, this has not been sufficient to absorb the natural increase in the labour force and the unemployment rate remains one of the worst in the Union at around 18 per cent [*Government of Ireland, 1993, Ch.1*].

JOBS VERSUS THE ENVIRONMENT

The persistently high levels of unemployment and emigration in Ireland have ensured that job creation has stayed at the top of the political agenda in recent years. Environmental concerns have, until recently, received much lower priority. Attitude surveys during the 1980s revealed Ireland to be consistently at or near the bottom of the league in terms of concern for the environment, especially when the issue was a trade-off between economic growth and environmental protection [*Whiteman, 1990*]. On the other hand, the country's late industrialization and urbanization meant that Ireland has not suffered anything like the same degree of environmental degradation as the more developed states. The issue for Ireland has mainly been to retain its relatively unpolluted environment, rather than the cleaning-up exercise which is the problem in the more industrialized countries.

Ireland is not without some significant environmental problems, but suffers less than many of its European partners, especially in the area of toxic waste. In general Ireland enjoys a high standard of air quality. Because of the prevailing westerly winds the country is not affected by trans-boundary air pollution as occurs in mainland Europe. Water pollution has been more problematic, arising from nutrient enrichment from agriculture, urban sewage and ineffective septic tanks. A recent review of water quality in Ireland for the period 1987–90 confirmed that serious water pollution is decreasing, but moderate pollution continues to increase [*Environmental Research Unit, 1993*].

The rapid industrialization and modernization of Ireland during the 1960s and 1970s was accompanied by new social movements, including ecology groups [*Baker, 1991*]. Since then there has been a steady growth in environmental awareness, with the establishment of Irish branches of international environmental groups such as Greenpeace, Friends of the Earth and the World Wide Fund for Nature (WWF), as well as a proliferation of locally based indigenous groups. In many cases the latter are loose alliances formed to oppose a particular project which is perceived to be harmful to the health or interests of the local community. In other cases local groups have merged to form permanent environmental watchdog groups for their locality or for specific issues. The Cork Environmental Alliance, for example, is an umbrella body for environmental groups in the Cork Harbour area where there is a high concentration of chemical and pharmaceutical industries owned by multi-national firms.

An issue of particular concern to environmental groups is toxic waste disposal. At present there is no facility for disposing of toxic waste, apart from a small number of 'in house' incinerators. The result is that waste must be exported, stored or disposed of in unsuitable facilities. The government is now committed to providing a national facility but there is considerable opposition from environmental groups, notably Greenpeace, to incineration, which is the option favoured by government officials and industry. In addition, the proposed location of the incineration site has been the subject of local NIMBY (not in my back yard) protest groups.

Overall, environmental groups have made an important contribution to raising environmental awareness among the public and encouraging critical public debates on a wide range of concerns. In 1989 the Green Party succeeded in wining its first seat in Dail Eireann (the lower house of Parliament) and retained a Dail seat, though in a different constituency, in the 1992 general election. The decision to include representatives of the Irish Women's Environmental Network and the National Youth Environmental Organization on the Advisory Committee to the newly established Environmental Protection Agency (EPA) reflects the success of the environmental lobby's claim to a role in the policy process. However, environmentalists have sometimes been met with hostility within their own community where they have been perceived as anti-development and anti-industry, and they have failed to win the support of trade unions and organized labour [*Baker, 1991: 63;* see also *Allen and Jones, 1990*, for an analysis of the ecology movement in Ireland].

A combination of legal obligations arising from EU environmental policy, as well as the desire to be seen as pro-active on environmental issues, has resulted in the introduction of an array of environmental initiatives by the Irish government in recent years. In 1990 the government published an *Environmental Action Programme* to provide a comprehensive and systematic framework for environmental protection in Ireland, and also established ENFO, the Environmental Information Service, to provide public access to information on the environment. An Environment Policy Research Centre has been established within the Economic and Social Research Institute to conduct research on the economic aspects of environment policy. In 1993 the national Environmental Protection Agency (EPA) came into operation in order to provide a more efficient system for managing environmental control (this body will be discussed in more detail below). The higher profile being given to environmental issues by the government is further reflected in the fact that the *National Development Plan 1994–1999* contains an entire chapter on 'Environmental Situation' [*Government of Ireland, 1993, Ch.12*] compared to the eight lines devoted to the environment in the 1989–93 Plan [*Government of Ireland, 1989: 15*]. The current Plan also pledges the government to a development strategy 'which fully respects the principle of environmental sustainability' [*Government of Ireland, 1993: 31*].

The government has been very anxious to promote Ireland's 'green' image in order to capitalize on the swing in consumer taste, especially in the tourism and food industries. A recent government-commissioned report ('The Culliton Report') on industrial policy for the 1990s stated that:

> Far from dragging our heels in the international movement for environmental protection, it could be in Ireland's interest, even

from the industrial viewpoint, to seek out opportunities for advancing ahead of other countries in the interests of promoting this [green] image and in positioning Irish industry for coping with ever-tougher regulations in the face of what seems an inexorable trend [*Government of Ireland, 1992: 33*].

The Report, however, goes on to stress the need to ensure that concern for the environment is not distorted by anti-industrial and anti-employment attitudes, reflecting the continuing dilemma of jobs versus the environment for a country with an unacceptably high level of unemployment.

ADMINISTRATIVE ELITES AND EU POLICY STYLE

Although Irish officials are now well socialized in the style of EU policy-making, there is a marked contrast between the relatively informal and loose institutional structure of Irish public administration and the more formal and deliberate approach adopted in continental Europe which is based on administrative law tradition. Irish policy style is fashioned on British administrative culture. The characteristic pragmatic approach results in a policy style which is reactive rather than pro-active. The small size of the Irish bureaucracy has made it very difficult for officials to keep on top of the impressive array of environmental policy measures emanating from the national government, the EU and international organizations in recent years. There is not the same degree of specialization in the Irish public service that is normally found in larger administrations. Individual officials are responsible for a much wider range of policies and the same official is usually responsible for both the negotiation of EU legislation and its incorporation into Irish law. On the positive side, the small scale means that the system is much less bureaucratic than in many other European states and there are fewer interests to aggregate and articulate [*Laffan, 1989*].

There is no evident hostility towards the involvement of the EU in a wide range of domestic policy areas. EU laws are regarded as legitimate and authoritative and given the same status as national laws [*Laffan* et al., *1988: 408*]. A combination of Ireland's small size and the desire of the Irish to be seen as 'good Europeans', means that Irish government ministers and officials are generally supportive of EU policies and are reluctant to go against the trend at the negotiation table. The desire to promote Ireland's 'green' image, by being seen as pro-active on environmental issues, means that the government has to support measures which are sometimes of marginal relevance to Ireland. Many EU directives are time-consuming and costly to implement and often require sophisticated monitoring equipment and procedures which place a heavy burden on a small public sector already under strain.

Government officials desire to apply faithfully the spirit of EU directives usually results in the verbatim transposition of directives into Irish law. This is not peculiar to Ireland; according to Schaefer [*1991: 110*], member-states seek to avoid being cited before the European Court of Justice for inadequate transposition of EU law, and so they tend simply to transpose the text verbatim into national law. 'Consequently, the major purpose of a Directive – namely to use the instruments appropriate to the Member States' legal and administrative systems in order to realize the objectives of a Directive – is violated and actual application is made more difficult' [*Schaefer, 1991: 110*].

The bulk of EU environmental legislation is in the form of directives which merely specify the results to be achieved but leave the choice of form and method to member-states. In Ireland EU directives are most commonly transposed into national law as regulations made under the 1972 EC Act, or more specific acts where appropriate. This is generally much less time consuming than enacting primary legislation but usually involves some element of consultation and inter-departmental co-ordination, which may take more time than is provided for in the directive. EU environmental directives are mostly technical in nature and new primary legislation is not normally considered necessary. However, the government sometimes consolidates existing disparate national and EU legislation into a new comprehensive single act, for example, the Air Pollution Act, 1987.

There are disadvantages associated with implementing EU directives by means of regulations. According to Scannell [*1990a: 100–1*], all of the Irish regulations implementing EU waste directives are defective. For example, only summary offences can be created under the 1972 EC Act, so the maximum penalty for offences created by regulations made thereunder is only IR£1,000 and/or six months' imprisonment. This can lead to anomalies. For example, a person who dumps cadmium illegally on land is liable to this maximum penalty. If the substance is dumped at sea, the penalty is an unlimited fine and/or 5 years' imprisonment. Another disadvantage of the use of regulations is that the possibility of securing the objectives of the directives by other more effective means than mere criminal sanction are rarely provided for. The Union of Professional and Technical Civil Servants (UPTCS) in Ireland points out that many sections of the 1976 Wildlife Act and related legislation are prohibitive in form and rely for their successful implementation on detection and enforcement, which are subject to a range of special problems, including a general public disregard for legal obligations, a low level of public awareness of conservation and the law, inefficient and over-centralized prosecution procedures and ludicrously low penalties for offences [*UPTCS, 1987: 31*]. Moreover, because regulations are made by government ministers more or less in camera, there is normally no public debate in Parliament or the media.

ADMINISTRATIVE STRUCTURES FOR ENVIRONMENTAL MANAGEMENT

In 1977 the Department of Local Government was re-named the Department of the Environment to reflect the growing importance of environmental issues. The Department was given overall responsibility for the formulation and co-ordination of environmental policy and for promoting the protection and improvement of the environment. The Department has responsibility for the negotiation of EU environmental directives and for ensuring their implementation. Apart from environmental issues, the Department's other functions include roads, housing, sanitary services and the supervision of local authorities.

Other government departments with a significant environmental interest are the Department of the Marine and the Office of Public Works, which has responsibility for the Wildlife Service. These departments retain a central control function for environmental issues intrinsic to their sectoral remit which require a national approach. Oil and mineral development is also handled centrally by the Department of Transport, Energy and Communication. Ultimately, each government department has responsibility for ensuring that the environmental impact of their policies does not conflict with national and EU standards.

Implementation of environmental policy which comes within the remit of the Department of the Environment, including EU directives, falls mainly to the local authorities. Local authorities are responsible for implementing most of the EU directives relating to water, waste and air.

ROLE AND CAPACITY OF LOCAL AUTHORITIES IN THE AREA OF ENVIRONMENTAL CONTROL

EU environmental measures impact on local authorities in two ways:

i) in setting standards for the operation of their own core functions in relation to water supply, sewage treatment and waste disposal;
ii) in their role as authorities responsible for the implementation and monitoring of a wide range of environmental measures.

Local government in Ireland has no constitutional basis, but rather local authorities have their origins in, and derive their powers from, Acts of the Oireachtas (Parliament). They have no general responsibility in matters of local interest and may be held to be acting *ultra vires* – outside their powers – if they perform functions not specifically laid down by law. The main local authority is the county council, of which there are 27, one for each of the 26 counties that comprise the Irish Republic, but two for County Tipperary for historical reasons. There are also county borough corporations in the five largest cities with the same powers as county councils. These 32 local authorities comprise the core of the local administrative system in terms of powers, functions and finance. The term 'local authority', as used in this chapter, refers to these 32 authorities. There is a lower tier of urban districts and a small number of Town Commissioners, and in January 1994 a new tier of regional authorities was established.

When contrasted with other European countries, the range of functions performed by Irish local authorities is very limited and mainly concerned with the physical environment, specifically with planning and development and environmental management and control. They have effectively no involvement in major policy areas like education, agriculture, social welfare, police, public transport and public utilities. Moreover, the exemption of agriculture and forestry from planning controls hampers the formulation of an integrated and comprehensive environment protection policy at the local level, especially since the farming sector is one of the main sources of pollution in Ireland.

The functions of local authorities are classified as 'reserved' and 'executive'. Reserved functions cover major policy issues, finance and legislation. Anything that is not a reserved function is deemed to be an executive function and falls within the responsibility of the manager, a non-political career official. Environmental management is, for the most part, an executive function. Political control is exercised only to the extent that elected members of the local authority make provision for environmental action in the annual estimates and adopt the five-year development plans for their areas.

Relationships between local authorities and central government are regulated through the Department of the Environment, which exercises financial, administrative and technical controls over their affairs. The Department does not normally consult local authorities on policy issues which impact on their operation, including EU environmental policy. Because of their statutory status, local authorities are not regarded as interest groups to be consulted on policy issues which affect them, and there is no evidence to suggest that their expertise as implementing agencies is reflected in the negotiations on new EU directives [*Hart, 1990: 36*]. Rather, local authorities are simply informed of new administrative practices required of them arising from policy changes or new legislation.

As the 'parent' Department, the Department of the Environment is the main channel of information to the local authorities on EU issues. Hart [*1990*] identifies a number of problems with this arrangement from a local authority perspective:

a) Officials in the Department do not necessarily have the same perspective on implementation problems as the local authorities;
b) The Department is relatively peripheral to EU negotiations except in areas for which it is specifically responsible, notably environmental protection. Other government departments play a more pivotal role but local authorities have no access to these except indirectly via the Department of the Environment;
c) The Department handles EU dossiers on a functional basis and individual sections liaise with their functional counterparts in the local authorities. The net effect is that local authorities do not have a centrally co-ordinated perspective on relevant EU policies.

In an analysis of the administrative and technical arrangements for environmental management in Ireland [*Leech, 1989*], only ten local authorities were found to have a separate section dealing specifically with environmental control. In most authorities the function was part of the sanitary services department. This is in marked contrast to the planning and development function for which the local authorities have a more rational and unified structure.

Resources of Local Authorities

The impressive array of environmental legislation and measures which have emanated from national government, the EC/EU and international organizations in recent years has not been accompanied by a corresponding development of institutional arrangements to implement and manage these measures. As EC directives came on-stream, the trend was to add these on to the existing functions of local authorities without adequate consideration of additional staffing and financial requirements.

a) *Staff*: The predominant discipline found in local authorities on the technical side is the civil engineer, 'whose training in water treatment and structures allow him to embrace the skills of water pollution control and waste management' [*Leech, 1989: 190*]. The technical aspect of environmental management is almost exclusively the reserve of engineers. An embargo on public sector recruitment during the 1980s, at a time when new demands were being placed on the local authorities, meant that the new technical expertise, especially chemists and biologists, required to conduct quite sophisticated environmental monitoring and control measures, were not available to the local authorities.

b) *Managerial Skills*: A recent Report on the training and development implications for local authorities arising from the completion of the Single European Market concluded that:

> One of the problems with the implementation of environmental measures at local level has been the lack of a co-ordinated approach at management level to the combined impact of directives. The implementation of directives has been the reserve of technical staff with the necessary expertise to fulfil the local authority's obligations. Insufficient attention has been given to actively managing environmental protection locally. One of the training needs emerging from this analysis is training in environmental management for senior staff in local authorities [Hart, 1990: 14–15].

The standard annual programme which local authorities implement encompasses a variety of over-lapping, multi-annual programmes and directives with different timescales and deadlines including:

i) a five-year Structural Fund Operational Programme for water and sanitary services;
ii) a three-year Operational Programme under ENVIREG which deals with sewage treatment in coastal areas;
iii) a ten-year National Environment Action Programme [Tallon, 1992].

It is the function of senior management in the local authorities to manage these programmes in an integrated manner, making the most efficient use of available resources. The dominance of the engineer on the technical staff has implications for the approach to environmental management adopted by the local authorities. Irish local authorities have been likened to major engineering firms. There has been much criticism of the priority given to infrastructure development in programmes funded by EU Structural Funds [Meldon, 1992]. This is partly a limitation of the rules of the Structural Funds themselves which require investments to show an economic return. The Fifth Action Programme on the Environment links economic development to the principle of sustainable development. Environmental protection is, however, easier to value than to price. As Tallon [1992: 12] asks, 'Pollution free waters are the raw material of many tourism activities but what economic weight, directly or indirectly, should they be given in relation to tourism growth?' One needs to take account of all the costs and benefits of actions and non-actions to determine priorities where benefits are highest; that is, one needs to have an overall management strategy. However, the narrow functional remit of local authorities in Ireland, together with the fact that they have a limited input into the policy process, including the National Development Plan for Structural Funds, means that their policy scope and freedom of manoeuvre for environmental management is very restricted.

c) *Organizational Capacity*: The lack of resources and organizational capacity within the local authorities is quite evident from the range of environmental management functions for which they rely on outside expertise. Leech [*1989: 207–10*] found that while waste management plans were carried out almost entirely by the local authorities' own staff, almost 80 per cent of the input into the drafting of water quality management plans were carried out by a semi-state research agency, An Foras Forbartha (now the Environmental Research Unit). Air Pollution and air quality monitoring is generally dealt with on an agency basis by health inspectors employed by the Health Boards which are outside the local authority structure.

Local authorities have their own laboratory facilities but also use the expertise of chemists attached to the three state Regional Laboratories, each of which serves several local authority areas. The national science and technology agency, EOLAS (incorporating the former Institute for Industrial Research and Standards and the National Board for Science and Technology) is available for more complex analyses.

Ideally, local authorities should have within their own structures the resources to develop and fully understand the technical implications of the control measures for which they have responsibility. However, in his evaluation of the administrative and technical arrangements for environmental management in Ireland, Leech concluded that:

> When the range of expertise required is considered covering the broad area of the aquatic environment, toxicity, ecotoxicology, assimilative capacity, standards, control technology, air pollutants, acid rain, toxic wastes and chemicals, air and water quality modelling, etc., it is evident that few local authorities employ or could afford to employ the broad range of expertise required [*Leech, 1989: 412*].

A recent report on *The Impact of Planning, Licensing and Environmental Issues on Industrial Development* said that the lack of adequate in-house resources to handle complex industrial projects leaves local authorities in a particularly weak situation for early consultation, particularly where several locations are being considered by the companies concerned and the Industrial Development Authority [*Benson* et al., *1992: 29*].

Conflict of Interests

The authority of local government in environmental control has been compromised by the fact that local authorities are, themselves, major polluters. This has led to a widespread disregard for pollution laws. Ultimately, there is a conflict of interest between the developmental role of local authorities and the control functions assigned to them under environmental legislation. Local authorities act as both gamekeepers and poachers in respect of water quality and waste disposal since, under existing arrangements, they are responsible for the environmental impact and control of their own operations.

Overall, implementation and enforcement are much stricter for the private sector than for the public sector. Scannell [*1990: 102*] gives the example of the EC (Waste) Regulation 1979 (implementing directive 75/442/EEC), which does not mention the obligations which the directive imposes to promote the beneficial uses of waste, carry out periodic inspections of waste disposal facilities, ensure respect for the 'polluter pays principle' and forward periodic situation reports to the Commission. All these obligations fall on the public sector. Likewise the EC (Toxic and Dangerous) Waste Regulations, 1982 (implementing directive 78/319/EEC) do not refer to obligations to take appropriate steps, as a matter of priority, to encourage the prevention of this waste, its processing and recycling. Nor does it mention the requirement to make periodic reports to the EC. Even more perplexing for Scannell is the complete failure to subject many local authorities disposing of toxic and dangerous waste to the permit system required by the directive. Article 9 of the directive requires *all* establishments storing, treating and depositing toxic and dangerous waste to obtain permits from the appropriate authority. The 1982 regulations do not require local authorities, the bodies most likely to accept dangerous waste, to obtain these permits. 'Thus there are no provisions in Irish domestic law prohibiting sanitary authorities from directly discharging List 1 dangerous substances to ground-water or providing for the authorization of sanitary authorities in respect of indirect discharges of these substances to ground-water' [*Scannell, 1990: 86*].

A recent report by An Taisce (The National Trust for Ireland) points out that in the Operational Programme for Water and Sanitary Services, for which the local authorities are responsible, there is no mention of the need for the local authorities to comply with directive 85/337/EEC on Environmental Impact Assessment [*Meldon, 1992: 41*].

NATIONAL RESPONSE: THE ENVIRONMENTAL PROTECTION AGENCY

In 1991 the government decided to establish a national Environmental Protection Agency (EPA) in recognition of the complexity and scale of environmental measures which had come into force over the previous few years, and the resultant pressure which had been put on the limited resources and expertise of local authorities. It was also an attempt to bring about a more rational and effective system for managing environmental control and a greater degree of integration and co-ordination in planning and licensing matters. According to Benson et al. [*1992: 42–3*] the need for the EPA derived from:

a) the acknowledged need for clear national standards and their enforcement in a consistent and even-handed manner;
b) the inadequate and uneven distribution of resources in the present planning system;
c) the inadequate performance and lack of credibility of some local authorities in their environmental policing role;
d) the need for a central environmental body to advise government and act as a focus in its relationship with the European Environmental Agency.

The EPA, which became operational in 1993, operates mainly in the following ways:

a) Direct action: responsibility for a new integrated licensing and control system for environmental management; preparation of guidelines on Environmental Impact Statements; specification of environmental quality objectives practices; maintenance of databases on the quality of the environment, to which the public will have access; and the publication of reports on environmental issues.
b) Adivsory and support functions: providing advice and technical support to public sector agencies, including local authorities, on environmental management.
c) Supervisory functions: the EPA will supervise the performance by local authorities of their statutory environmental functions.

In a speech outlining the role of the EPA, the then Minister for the Environment emphasized that it should not be seen as anti-industry and that it will have regard for ensuring a proper balance between environmental protection and the need to provide for infrastructural, economic and social progress and promote sustainable development [*Harney, 1991: 31*]. Greenpeace has criticized the adoption of the BATNEEC (best available technology not entailing excessive costs) principle by the EPA. It is also critical of the absence of any provision for the appeal of decisions by the EPA to a third party; instead there is simply a procedure for lodging objections to the granting of licences [*Greenpeace, 1993*]. The Benson Report on the impact of licensing and environmental issues on industrial development also criticized the absence of an appeal system, other than through the courts. It points out the inconsistency in allowing appeals in the case of licences issued for non-scheduled activities or activities below certain thresholds: 'This is clearly illogical and inconsistent and is likely to encourage applications marginally below the threshold values' [*Benson et al., 1992: 34*].

The decision to establish the EPA was generally welcomed and regarded as probably the most expedient response to addressing the glaring inadequacies in the existing system for managing the environment. There is, nevertheless, regret among some that this response represents a further erosion of the ever-decreasing powers of local authorities in Ireland. Over the years, as new demands have been made on government, the trend has inexorably been to establish new single function executive agencies at national or regional level rather than to upgrade the capacity of local authorities and devolve functions to them. This has inevitably resulted in the failure of local authorities to perform adequately their monitoring functions in relation to the environment and, indeed, to the fact they have themselves been significant sources of environmental pollution. Numerous instances in recent years of their failure or inability to monitor and implement the conditions of their own pollution licences has led to the breakdown of public confidence in their capacity for environmental management and, indeed, in the priority which they accord to this function.

PERIPHERALITY – A NECESSARY BUT NOT SUFFICIENT EXPLANATION

In the less developed regions of the EU, the costs of environmental protection and the value of sustainability have to be assessed against competing and critical demands for funding. The Irish government has decided to adopt a pro-active role in environmental protection and has taken several initiatives to promote a more open and rational system of environmental management, notably the establishment of the EPA. Nonetheless, the sheer volume of environmental protection measures to be enacted and implemented has put a huge burden on the public administration system at the central and local levels. Consequently, there is generally no time or expertise to evaluate the scope of EU directives and to devise the most appropriate ways of implementing them. According to Scannell [*1990b: 89*], a mixture of administrative convenience, legislative inertia and lack of guidance has led to a situation in which some EU directives are being 'over-applied'. She cites the example of directives 78/659/EEC and 79/923/EEC on water quality standards for waters supporting fish life, which are being applied automatically by some local authorities to *all* waters into which trade (but not sewage) effluents are discharged, irrespective of the condition of the receiving water, and even though the directives permit member-states to derogate from their parameters. This experience:

> ... illustrates the danger of introducing EC standards without proper guidance and debate on when they ought to be applied in Ireland. It is apparently more convenient for more unqualified administrators to impose standards which seem to have a legislative imprimatur than to commission the expertise required to assess the impact of an individual licence application on receiving waters properly [*Scannell, 1990: 89*].

According to Schaefer [*1991: 111*], addressing the differences in the administrative capacities and competences of member states is a priority if the EU is to adopt a leading role on the environment in the international arena. He maintains that it is not sufficient for those member states which have the know-how and financial resources to support the necessary administrative capacity simply to demand that the economically weaker member states attain the same performance. The EU has responded to this problem with a variety of policy instruments aimed at up-grading the infrastructural, technical and administrative capacity of the peripheral countries through the Structural and Cohesion Funds. Ireland has secured substantial financial assistance under a variety of EU programmes for a wide range of projects:

a) the Operational Programme for Water, Sanitary and Other Local Services has secured EU grant assistance of IR£110 million over the period 1989–93;
b) the ENVIREG Operational Programme, which caters for the provision of secondary treatment and sludge management/disposal facilities for coastal towns, received an EU aid contribution of IR£23 million;
c) environmental measures under the INTERREG initiative for border areas attracted EU funding of IR£6.9 million;
d) Ireland is also participating in a number of other EU initiatives on the environment, notably ENVIRONET (a trans-European data net-

work on the environment) and LIFE (a development of innovative techniques or processes for solving environmental problems) [*Department of the Environment, 1993*].

While financial transfers from the EU under the forthcoming Structural and Cohesion Funds will undoubtedly enable Ireland and other peripheral member-states to improve further their capacity to deal with environmental problems, capital investment should only be directed towards those problems which are not otherwise solvable. Often the ready availability of finance from the EU means that other less costly solutions are not considered. The absence of an overview of environmental management is compounded where responsibilities are fragmented and uncoordinated. In the case of Irish local authorities, their narrow functional remit and resources limit the options available to them in dealing with environmental problems at the local level. Thus they tend to deal with water pollution through end of pipe solutions which generally involve infrastructural investment (for which they can get Structural Funding), rather than at source by, for example, re-educating the farmers who are major offenders but over whom the local authorities have no responsibility.

Sound environmental management requires a combination of infrastructural investment, rational administrative structures and policy coordination at the local, regional, national and EU levels. These are certainly requirements which impinge more on the peripheral countries of the EU and which can be partly solved by the transfer of financial assistance and technical know-how. Ultimately, however, there is a vital ingredient which can only be supplied by the individual member-states themselves and this is a genuine commitment to tackle environmental degradation and to accord it a high priority on the political agenda, not just in respect of enacting laws and establishing institutions, but, more importantly, in implementing those laws and imposing sanctions to ensure compliance. This is a much more complex issue for member-states and one which is bound up with historical and cultural traditions as much as with socio-economic circumstances. Thus, as Coakley points out:

> For all their deferences to strong leaders, the Irish are not noted for their compliance with laws and regulations . . . . Thus many aspects of deviant behaviour, such as traffic violations or tax evasion, enjoy a considerable degree of public tolerance and are seen as legitimate contests between the individual and the agents of the state . . . . Irish people defer to authority collectively in principle, while reserving the right individually to frustrate it [*Coakley, 1993: 38*].

Ultimately, the commitment to the environment must be valued for its own sake and not arrived at after an assessment of the economic gains accruing to the economy from espousing 'green' ideology. The Irish government has recognized the potential economic advantages for the food and tourism industries in promoting a clean environment. However, as O'Donnell points out, while this is accurate it, in turn, reflects a dependent mentality that may be part of Ireland's economic problem [*O'Donnell, 1993: 34*]. The literature on international competitive advantage suggests that in general the process works the other way round: 'That is, countries tend to develop an international competitive advantage in activities which they perform to a high standard *for themselves*' [*O'Donnell, 1993: 35*].

NOTES

1. The term 'Ireland' is used in this chapter to refer to the Republic of Ireland, and does not include Northern Ireland.

REFERENCES

Allen, R. and T. Jones, 1990, *Guests of the Nation – People of Ireland versus the Multinationals* (London: Earthscan).
Baker, S., 1991, 'The Evolution of the Irish Ecology Movement', in W. Rüdig (ed.), *Green Politics One* (Edinburgh: University Press).
Benson, F.L. and Associates, Matheson Ormsby and Prentice and Ove Arup, 1992, *The Impact of Planning, Licensing and Environmental issues on Industrial Development: A Report to the Industrial Policy Review Group* (Dublin: Stationery Office).
Coakley, J., 1993, 'Society and Political Culture', in J. Coakley and M. Gallagher (eds.), *Politics in the Republic of Ireland*, 2nd edition (Dublin and Limerick: Folens and PSAI Press).
Department of the Environment, 1993, *Environmental Bulletin* (Jan.–June) (Dublin: Department of the Environment).
EC Commission, 1992, *Towards Sustainability: A European Community Programme of Policy and Action in relation to the Environment and Sustainable Development*, Com(92) 23 final, 27 March.
Environmental Research Unit, 1993, *Water Quality in Ireland 1987–1990* (Dublin: Environmental Research Unit).
Government of Ireland, 1989, *Ireland: National Development Plan 1989–1993* (Dublin: Stationery Office).
Government of Ireland, 1992, *A Time for Change: Industrial Policy for the 1990s: Report of the Industrial Policy Review Group* ('Culliton Report') (Dublin: Stationery Office).
Government of Ireland, 1993, *Ireland: National Development Plan 1994–1999* (Dublin: Stationery Office).
Greenpeace, 1993, *Stop Legal Pollution* (Dublin: Greenpeace Ireland).
Harney, M., 1991, 'The Irish Environmental Protection Agency', in J. Feehan (ed.), *Environment and Development in Ireland* (Dublin: The Environmental Institute).
Hart, J., 1990, *Completion of the Single European Market: Training and Development Implications for Local Government*, unpublished report prepared by Laois County Council.

Laffan, B., M. Manning and P.T. Kelly, 1988, 'Ireland', in H. Siedentopt and J. Ziller (eds.), *Making European Policies Work: The Implementation of Community Legislation in the Member States* (London: Sage).

Laffan, B., 1989, 'Putting European Law into Practice: The Irish Experience', in *Administration*, Vol.37, No.3.

Leech, B.C., 1989, *The Administrative and Technical Arrangements for Environmental Management in Ireland*, Vols.1 and 2, unpublished PhD dissertation, Trinity College, Dublin.

Meldon, J., 1992, *Structural Funds and the Environment: Problems and Prospects* (Dublin: An Taisce).

O'Donnell, R., 1993, *Ireland and Europe: challenges for a new century*, Policy Research Series Paper no. 17 (Dublin: Economic and Social Research Institute).

Scannell, Y., 1990a, 'Legislation and toxic Waste Disposal in Ireland', in *Environment Protection and the Impact of European Community Law*, Papers from the Joint Conference with the Incorporated Law Society of Ireland (Dublin: Irish Centre for European Law, Trinity College).

Scannell, Y., 1990b, 'Impact of EC Water Pollution Directives in Ireland', in *Environment Protection and the Impact of European Community Law*, Papers from the Joint Conference with the Incorporated Law Society of Ireland (Dublin: Irish Centre for European Law, Trinity College).

Schaefer, G.F., 1991, 'The Subsidiarity Principle and European Environmental Policy', in *Subsidiarity: The Challenge of Change: Proceedings of the Jacques Delors Colloquium* (Maastricht: European Institute of Public Administration).

Tallon, G., 1992, 'Water and Sanitary Services: Structural Funds and the Environment', unpublished paper delivered at training course for local authorities organized by the Institute of Public Administration, Dublin.

Union of Professional and Technical Civil Servants, 1987, *Our Natural Heritage: A Policy for National Conservation in Ireland* (Dublin: Brunswick Press).

Whiteman, D., 1990, 'The Progress and Potential of the Green Party in Ireland', *Irish Political Studies* Vol.5, pp.45–58.

National Environmental Policy-making in the European Framework: Spain, Greece and Italy in Comparison

GEOFFREY PRIDHAM

I. INTRODUCTION

The European integration process involves a complex set of interactions between the European Union (EU) institutions and the political systems of member-states. According to an early study of EC policy-making:

> The Community process is not confined to what takes place within the formal framework of the Community institutions. Rather it embraces a network of relationships and contacts among national policy-makers in the different member-states, both directly through involvement in the Community arena and indirectly as that arena impinges on national policy processes ... the Community process ... can be analyzed only as the tip of a much larger iceberg formed by the domestic contexts that set constraints on each member government. In addition national policy-makers are caught up in other kinds of transnational activities including other international agencies and a variety of informal links both multilateral and bilateral. These arenas of discussion sometimes complement or reinforce the Community process, but on occasion may complicate or undermine it [*Wallace, 1977: 33–4*].

Since this was written, the policy scope of the EU has expanded significantly so that it is now difficult to make any clear distinction between EU and national affairs. Environmental policy, where the EU has become more activist since the mid-1980s, is one important example of this change. During the same period national publics have, to differing degrees, become more sensitized to environmental questions. In this area there is therefore considerable potential for politicized interaction between EU institutions and national systems at different levels.

This chapter examines the contribution of EU policy pressures to the development of national policy in southern Europe. While it is supposed that the three southern member-states of Italy, Spain and Greece are broadly similar in their position on environmental questions – they are all, for instance, included in the category of 'cohesion countries' (recipients of the EU's new Cohesion Fund) – differences between them are also significant. These differences tend to be emphasized by the effect that Brussels has on the motivation and coherence of their environmental policy as well as on their policy style and implementation.

2. CONFRONTING SOUTHERN EUROPE OVER THE ENVIRONMENT: PROBLEMS AND ISSUES

The capacity of individual countries to respond to the challenge of ecological modernization and, more recently, sustainable development – an outlook that has come to influence EU policy-making – cannot be assessed without reference to national systemic factors. As Weale notes, cross-national difference here 'is deeply rooted in policy styles and organizational structures', comprising a mix of institutional and ideological factors, and as such 'is likely to be difficult to change' [*Weale, 1991: 21*].

The reputation of these southern countries for ineffectiveness and corruption, administrative lethargy and defective policy co-ordination clearly has profound implications for their capacity to respond to what Weale refers to as the 'new politics of pollution'. These are all countries in which the state has been over-developed, playing a dominant part in the economy. Furthermore, the prevalence of consumerist values in these recently modernized countries presents a powerful obstacle to environmental values. Moreover, traditional features of a country's culture may affect response to the environment, such as national pride in '*patrimonio*' – to use the term widely employed in both Italy and Spain – which is often idiosyncratic. An example of this would be the particular Greek concern for national monuments. To add to this, there is a strong pattern of localistic cultures in the South, introducing a territorial dimension to environmentalism. Public feelings might become sensitized over a location-specific environmental issue or event and exert pressure on parties and their leaders. Indeed, there are signs of environmentalist values gaining ground in southern Europe in recent times.

In terms of broad characteristics, there is a case for considering member-states in the Mediterranean region as a group – bearing in mind climate, physical features and, of course, the common problems relating to the Mediterranean Sea. The physical peculiarities of that sea, including its low volume of water, make it especially vulnerable to environmental damage through pollution – such as industrial and chemical waste, domestic sewage and oil pollution. The total human population in the area has increased and is increasing rapidly. Around the coast there are 537 cities with populations of 10,000 or more, 70 per cent of them in EU countries (*The Economist*, 21 Dec. 1991). Italy, Spain and Greece are also countries which have been modernizing fast. Greece, for example, has experienced rapid urbanization (notably in the Athens area) and the expansion of tourism (elsewhere in Greece), which have combined with weak infrastructure and the failure of planning controls to produce considerable disruption [*OECD, 1983: 20, 30, 120.*].

Destructive consequences for the environment have been exacerbated through EU membership, notably by the application of Structural Funds to the south. The Commission's Task Force report on the Single Market and the Environment (1989) noted the severe problems of environmental degradation in southern Europe and expressed alarm over the environmental consequences of the Single Market for the region, especially given the predicted growth in tourism [*EC Commission, 1989: 4.7–4.8, 3.21–3.25*]. There is a general sense that the southern member-states are markedly behind the most advanced Western countries in the environmental field. The Ministry of the Environment in Rome has admitted that 'the Italian government certainly finds itself, in certain respects, behind compared with the principal Western countries, which many years before Italy created structures assigned to the management of environmental problems in their respective public administrations' [*Ministero dell'Ambiente, 1989: 1.1–1.2*].

At the same time, it is important not to ignore ecological diversity in the region. This warns against too rigid a distinction between countries from the south and the north and requires that we take account of cross-national, not to mention intra-national, differences within southern Europe. These differences may be summarized as follows:

i) *Environmental administration*: there are differences at both national and sub-national levels. These countries follow different models, with Italy opting for a small ministry responsible for environmental policy while Spain and Greece have larger ministries comprising other areas (notably public works) as well as the environment. Accordingly, conflict of policy interests, notably between development projects and the environment, are in the first instance (Italy) inter-ministerial and in the other cases intra-ministerial. The three countries also vary sub-nationally in that Greece has centralized control over environmental affairs, while Spain and Italy respectively have quasi-federal and devolved structures.

ii) *Length of EU membership*: the three countries joined the Union at different stages – Italy was a founder member from the early 1950s, Greece acceded in 1981 and Spain in 1986. Italy was therefore involved in shaping EC environment policy from the start in the early 1970s, while the other two countries joined when the EC was expanding its activity in this area.

iii) *Implementation deficit*: while the southern states have a reputation for being tardy in transposing EU directives into national law (though they are not unique in this respect), they have shown some differences. Italy has by far the worst record of the three (and the worst in the EU), while Spain's record is among the best. There is often, however, a big gap between enactment and enforcement of EU legislation: Greece has a much better record than Italy on the first count but on the second is notoriously ineffective.

iv) *Socio-economic development*: these countries developed in different periods and in somewhat different ways. Italy urbanized earlier in the post-war period, while Spain and Greece did so mainly from the 1960s. Spain and especially Italy are generally more industrialized than Greece, whose industry is rather localized – especially in the Athens area. These differences have predictably influenced the nature and degree of environmental impacts in these countries.

v) *Environmental awareness*: it is possible to note some broad differences between the three countries. For instance, the EC Commission Task Force report of 1989 recorded that Italy and Greece were high up among EU states in perceiving environmental problems as urgent. It also noted an above-average level of 'don't knows' in southern Europe, except for Italy, reflecting possibly the limited degree of environmental education [*EC Commission, 1989: 1.9–1.10*].

vi) *Regional differences*: there is a pronounced divide between north and south in Italy as to degrees of industrialization and urbanization, not to mention culture; in Spain, there are regional differences, with Catalonia and the Basque Country having a high degree of industrialization not met elsewhere; while in Greece the main divide is between the two main cities (Athens and Thessaloniki) with high population density and the rest of Greece. Predictably, apart from determining the incidence of environmental problems, these differences have influenced the response to environmental issues.

vii) *Differences of specific environmental concern*: Spain has a particular interest in desertification and soil erosion, not to mention natural habitats; Italy and Greece are rather more concerned about the state of coastal water quality, with the latter generally more successful in maintaining such quality. Both countries are increasingly concerned about urban air pollution. This is especially true in the north and the centre of Italy, while in Greece, Athens forms the main focus of attention in matters relating to air quality.

The similarities and differences among the three southern European countries are elaborated in the following sections. In particular, changes arising from new environmental pressures at the national and European levels will be identified. We look in turn at national governmental structures and whether these facilitate or hinder environmental policy, at southern European responses to EU policy on the environment and then at different actors and influences under the heading of 'society'. It is hypothesized that the first (national governments) are more likely to give priority to environmental policy and to embrace new directions when there is combined pressure from the second (the EU) and the third (society).

3. THE STATE DIMENSION: INSTITUTIONAL LETHARGY OR POLICY ADAPTATION?

Policy Structures

In a recent study, Lopez Bustos presents a range of organizational models for environmental policy [*Lopez Bustos, 1992: Ch.7*]. These include: the concentration of competences in an environmental ministry; partial concentration of such competences in the same (with the environment ministry playing a co-ordinating role); partial clustering of competences in an already established (non-environment) ministry; the dispersal of competence between various ministries; and, hyper-sectorialization, whereby each ministry incorporates an environmental division. Italy, Spain and Greece fall into the second and third groups in terms of policy structures. Various points about these structural distinctions help to put the southern European countries into comparative perspective.

Firstly, given the relatively recent importance of environmental policy, there is a general difficulty in introducing effective environmental administration into traditional bureaucracies as logically that involves some radical restructuring of ministerial responsibilities; and this is bound to encounter bureaucratic resistance because of established interests in the government machinery. Of the three, the most fragmented is the Greek case. The Ministry of Environment, Physical Planning and Public Works in Athens has a number of limited functions, but many other ministries have environmental functions too: Merchant Marine covers protection of the marine environment; Health tests sea water quality and classifies beaches; Agriculture is responsible for protecting forests and monitors rivers; and Transport monitors car emissions [*Commission of the EC, DG XI, 1993: 57–8, 66*]. Institutional fragmentation is to some degree inevitable in a policy area that is itself cross-sectoral, but this is not, however, peculiar to southern Europe.

Secondly, these problems can in part be neutralized by effective co-ordination at both horizontal (inter-ministerial) and vertical (centre-periphery) levels, but it is here that the southern European countries are notably deficient. Ministerial rivalry and bureaucratic lethargy have proved powerful, as has the weakness of efficiency values and professional competence (as distinct from clientelistic practices). For instance, the Italian Ministry of the Environment attempted to introduce a national agency for environmental protection, 'with offices located territorially, and to which would be attributed the chief task of proposal, planning, control and verification of technical environmental standards' [*Ministero dell'Ambiente, 1992a: 74*]. But, within a year of this being announced, the draft law was blocked because of objections from other ministries involved, especially the Ministry of Industry which was strongly opposed to its creation [*Ministero dell'Ambiente, 1992a: 74*]. As to vertical co-ordination, this is crucial in such systems as the Spanish and Italian where environmental administration is not centralized. But inadequate procedures have magnified considerably the problem of vertical policy co-ordination. This has only begun to change in Italy with efforts by the Environment Ministry to centralize procedures through control over finance and impact assessment and the use of triennial plans [*Lewanski, 1993: 22–4*].

Thirdly, the different organizational models and their functioning raise the related problem of political weight. Thus, Italy has concentrated some tasks in the Ministry of the Environment, but its staff is extremely small compared with the larger and older ministries. It has only 300 employees, while Interior has 140,000, Labour 16,000, and Agriculture 4000 [*Hine, 1993: 232*]. Its report on its first five years of existence complained of its not being given an effective administrative structure, in particular with respect to adequate personnel, territorial articulation (field agencies) and resources [*Ministero dell'Ambiente, 1992a: 7–8*]. It also has logistical problems as its offices are located in four different parts of Rome. In Spain, national environmental responsibilities are partially concentrated in the large and powerful Ministry of Public Works and Transport, but the former have usually been subordinate to the concerns highlighted in the Ministry's title.

The outcome is that problems of environmental management experienced by the southern European countries are not as a whole unique, but there are special difficulties, particularly in relation to administrative procedure and competence. At the same time, the EU has created a consistent pressure on these countries to consider and even implement certain new procedures. These point to a possible new trend in institutional adaptation. For instance, Spain has, as the most recent entrant of the three, felt under compulsion to consider environmental planning for the first time and also the creation of a proper Ministry of the Environment [*Ministerio de Obras Publicas, 1990: Ch.9*].

Policy Style and Policy Patterns

While institutional history accounts partly for the problem of introducing effective environmental management, a low priority has often been accorded environmental policy by the governments of these countries, as reflected in their approach to problem-solving, policy outlook and degree of activism in this area. This is not entirely surprising since we are talking about countries – Italy is a partial exception – which have arrived late on the environmental policy scene, compared with countries in northern Europe. There have, however, been some recent signs of a more preventive or strategic approach, especially in Italy, but also at the regional level in Spain.

Evidence of low policy priority is either explicit, such as in public statements from key government leaders, or implicit, as in the priority given to policy which commonly conflicts with environmental concerns; in Spain's case, Gonzalez's dismissive approach to the environment and greater stress on the habitual Spanish preoccupation with water provision. This was on the occasion of the tenth anniversary of the Spanish Socialist Workers' Party's (PSOE) accession to power in 1992 and illustrated the line that had characterized his government over the previous decade. In Spain, the possibility of a new direction in environmental policy is also inhibited by a policy style that is closed and bureaucratic and without institutionalized channels for consultation with interest groups [*Aguilar, 1991: 1, 7*].

In Italy, the traditional policy approach has been dominated by response to crisis or emergency, leading invariably to a flurry of hasty legislation via decree. It is only in this *ad hoc* manner, responding to such temporary pressure, that bureaucratic lethargy has been overcome. It is indicative that Italian governments have invariably favoured detailed EU regulation of environmental matters which has then simply been incorporated into national law without undergoing the lengthy business of processing their own legislative proposals [*Rehbinder and Stewart, 1985: 140*].

Generally, these countries exhibit at best incrementalist rather than rationalist styles. Changes occur slowly and are usually of a minimal kind; but the absence of environmental policy strategy has usually meant traditional and especially economic concerns have remained predominant. Giorgio Ruffolo, Italian Minister of the Environment 1988–92, in his statement accompanying the first national report on the state of the environment in Italy (1989), spoke bluntly of his country's obsession with the gross national product and of 'basic cultural resistance to accepting the idea of sustainable development'. He goes on to argue that:

> Environmentalist policy is conceived still, to a large degree, as something external, peripheral and sectoral with respect to the production and consumption processes. Its actions are principally understood as *ex-post facto*, for repairing damage and reducing destructive and polluting effects. The concerns and resources of scientific and technological research are oriented predominantly towards progress in work productivity and product competitiveness . . . [*Ministero dell'Ambiente, 1989: Nota Aggiuntiva: 37*].

Italy has a high proportion of environmental laws, but, as Capria has noted, '. . . their production is strongly conditioned by the economic situation: only when market reasons and decisions of energy policy favour them and don't hinder their promulgation, then they are published' [*Capria, 1991: 13*]. Similarly, in Spain, the overriding concern with high unemployment has made it difficult to break the traditional view that environmental protection and employment generation stand in an antagonistic relationship to each other [*Ministerio de Obras Publicas, 1989a: 22*]. Such resistance to the ideas of sustainable development in policy thinking has been revealed also at the European level, where Spain has been virulently opposed to the idea of a carbon tax while Greece has resisted Commission proposals on chlorofluorocarbons (CFCs). Greek governments have placed a priority on acquiring EU resources for development, as from the Structural Funds, irrespective of their environmental consequences. Of the three countries, only Italy has shown a readiness of late to consider seriously policies promoting sustainability.

In these countries the main exception to the pattern of reactive response is to be found in the Italian attempt at the start of the 1990s to act strategically on environmental matters. Ruffolo's insistence on shifting to a preventive approach led to the development of three-year and even ten-year programmes to back up his intention. He also argued for increasing environmental expenditure, enlarging environmental legislation and for a more activist line by Italy at the international level to confront problems of ozone depletion and the greenhouse effect [*Ministero dell'Ambiente, 1992: preamble*]. The government has also begun to think about eco-taxes and to promote recycling schemes and voluntary agreements with major industrial groups [*Lewanski, 1993: 13*]. In southern Europe, such initiatives have been linked more to individual ministerial commitment than to a collective redirection of approach towards integrating environmental concerns into other ministerial portfolios. In the early 1980s and early

1990s the Greek environment ministers, Antonis Tritsis and Stefanos Manos, respectively, pushed for new environmental legislation, especially over air pollution in Athens. At the same time, they suffered from institutional constraints as did Ruffolo – above all, fragmented responsibility for the environment – so their activism did not produce permanent effects in terms of policy priority. The policy change in Italy may, however, be a new trend, linked as it has been to greater public concern.

At the sub-national level the idea of sustainable development, specifically in tourism, has, however, begun to affect policy thinking, albeit in a few regions rather than as a general phenomenon. This may suggest that the periphery in some cases is ahead of the centre. In Italy, the region of Emilia-Romagna has been in the forefront of pilot schemes to make the environment an integral part of land-use and industrial development planning [*Nanetti, 1990: 145–70*]. This is also true of Catalonia and Andalusia, with their more activist line over environmental affairs. Andalusia has one of the largest environmental administrations in Spain, including an environmental agency [*Commission of the EC, DGXI, 1993: 72*]. A notable case is the island of Majorca, where the regional administration has introduced new quality regulations and laws controlling traffic and the building of tourist facilities [*The Financial Times, 5 August 1992*]. Majorca's exclusive dependence on the tourist trade combined with a sensitivity to changing attitudes among up-market tourist families and a traditional appreciation of nature conservation have encouraged this policy line. It is now being used by the International Federation of Tour Operators (IFTO) as a model for upgrading the tourist industry in the Greek island of Rhodes as well as in Ireland (*The Financial Times*, 5 August 1992). According to the IFTO president, 'there is no altruism of any kind involved; it's absolutely straightforward – unless something is done we won't even have business' (*The European*, 30 July–2 August 1992). It goes without saying that environmental degradation, albeit in part a consequence of mass tourism, hits at the heart of the tourist industry's interests; and that in southern Europe there is a compelling need for tourism policies that respond to growing demands for sustainable development.

Policy Infrastructure

Policy facilities or resources of expertise may well strengthen the competence of policy-making in the environmental field. They include planning and monitoring mechanisms, efficiency of data collection and regularity of environmental information and the availability of expertise and environmental research. It is in this respect that the southern European countries are distinctly behind those of northern Europe.

Changes are beginning to occur at a basic level, such as the appearance of regular official reports on the national state of the environment. These have drawn on committees of experts both within ministries and those allied to national research councils. In Italy, the first such report was published by the Ministry of the Environment in 1989, with the second following – with a certain delay – in 1992. The Ministry of Health in Rome also now publishes detailed reports on the state of water quality at all Italian beaches; and the results – beach by beach – are reproduced in the press in the early summer. Clearly, that is an issue that arouses public interest. The Spanish Ministry of Public Works has, since 1984, published yearbooks on the state of the environment and legislation in addition to special volumes on such matters as environmental education and the principles of environmental law. In Greece, there have only been occasional reports. In 1983, the Organization for Economic Co-operation and Development (OECD) published a short though useful report on environmental policies there [*OECD, 1983*], but it has not been updated since; and Greece, like other countries, issued a somewhat generalized report for the Rio conference in 1992 [*Minister of Environment, 1991*].

The southern European countries have in this respect been following an international trend; but in other ways they are weak in policy infrastructure. Within the EU, they are usually the states with the least developed systems of planning control and with limited capacity for judging the potential damage done by investment in development projects (*The Economist*, 14 October 1989). This deficiency has been notorious in the case of Greece, where the lack of planning was highlighted by the controversy over the Prespa National Park, a pilot scheme for the Integrated Mediterranean Programmes which went wrong environmentally [*Baldock and Long, 1987: 17–19*]. Greece has some infrastructural mechanisms, such as the Athens Environment Pollution Control Programme (PERPA), the agency which daily tests air quality in the Athens area for the Environment Ministry and the monitoring stations of the Ministry of Merchant Marine which is responsible for the quality of sea water. Spanish laws on environmental protection have traditionally been marked by their lack of planning, although Madrid has increasingly come under pressure from the EU for changes in this respect [*del Carmen, 1986: 13*]. As the Ministry of Public Works has admitted, the EU has been crucial in launching the process of data collection on the environment as a necessary precondition for policy planning [*Ministerio de Obras Publicas, 1990: 11*].

Italy has a relatively extensive system of monitoring, but, as the Ministry's report of 1989 admitted, has problems with the reliability of estimates and completeness of information [*Ministero dell'Ambiente, 1989a: 152–3*]. Some improvement had occurred by the time of the 1992 report following a greater recognition of the importance of hard environmental information for public policy; but there remained insufficient expertise in public administration, according to the head of the environmental impact division:

> The analysis of the environment, whether for getting to know the Italy in which we live, or for defining the programmes for action and estimating their success, is not now considered an option, but is recognised – by all and in each institutional office – as a laborious necessity; however, we are not endowed with suitable personnel and the capacity for linking the world of public administration with the technical contribution of different indispensable disciplines [*Ministero dell'Ambiente, 1992: 8*].

As a recent survey of air pollution control in EU member-states showed, the establishment of a monitoring service in Italy was not followed by adequate financing, thus making implementation difficult. For example, no systematic sampling and analysis programme could be undertaken and the data collected was not sufficient for a complete picture [*Bennett, 1991: 32*]. In Spain, research and development is especially lacking in the environmental field, as a study sponsored by the Ministry of Industry has noted. It argued the need for Spanish scientific participation in the EU and other international organizations as a way of overcoming this infrastructural deficit [*Estevan, 1991: 518–19*].

In short, the southern European countries do not always fully deserve their reputation for backwardness in the environmental field. In some respects they are not entirely different from northern member-states. However, they stand somewhat apart with their limited professionalism and infrastructural facilities and, to some extent, in their traditional concern for the economic imperative. If changes have begun to occur, the European Union has been a significant although not sole or isolated influence; and it is to this we now turn.

4. THE EUROPEAN UNION DIMENSION: PRESSURE FOR POLICY CHANGE OR RECIPE FOR SYSTEM OVERLOAD?

The EU has become a real pressure for environmental policy change both in principle and in practice. In principle, this is evident from the Single European Act's emphasis on integrated pollution control and the preference for the precautionary principle and preventive action in the Maastricht Treaty. One study, looking at air quality standards, for instance, noted the 'radical challenge' EU environmental strategy could represent for national approaches to pollution control [*Elsom, 1987: 210–*

14]. At the same time, various features of EU policy and procedure have tended to exacerbate the problems for southern European countries in responding to EU environmental initiatives.

The stimulus to environmental legislation from Brussels has affected even Italy, a long-standing member-state. As Capria concluded, 'what distinguishes Italian legislation from others' legislation, above all in Northern and Central Europe, is the absolute prevalence of rules which owe their inspiration to the EC rather than the national level', for before the EC's first action programme Italian legislation on the environment was thin and 'first generational', meaning basic and inadequate [*Capria, 1991: 1*]. Spain, which joined more than a decade after EC legislation began, spent the first couple of years of membership enacting a whole backlog of EC environmental laws. Unlike Portugal, Spain did not request a delay in this enactment requirement because, basically for political reasons, it wished to be seen as participating fully in Europe [*Ministerio de Obras Publicas, 1989b: 245*]. For Spain, therefore, the effect of EC policy was more sudden and disruptive than for Italy or Greece.

Government leaders in the south have recognized that more adaptation is required of them than of their northern counterparts. Jose Borrell, Spanish Minister of Public Works, noted that the southern member-states including Spain 'will have to make a greater effort than those of the North' in view of their climatic characteristics and their tourist industries (*Pais Internacional*, 2 March 1992). His government's Industry Ministry acknowledged this was due to the 'lack of preventive and corrective measures in the last fifteen years'.[*Estevan, 1991: 183*]. The southern countries have, therefore, in recent years felt increasingly on the defensive over the growing pressure for environmental improvement in the EU. The debate over such issues has also tended to exacerbate the sense of a north/south dichotomy over environmental issues. As one Spanish official put it in reference to a dominance of particular northern concerns, this is 'a policy which has concentrated more on problems of industrial pollution and measures to combat it, than on programmes on the assessment, protection and recovery of soil, flora and fauna, and of making proper use of resources to avoid the progressive impoverishment and waste of nature' [*Baldock and Long, 1987: 59*].

For some time, contrary pressures have been evident between the legislative or standard-setting approach, urged by certain northern countries like Germany (concerned that their own advanced standards might be weakened at the European level) and the tendency of other member-states to look to the EU for resources in dealing with particular environmental hazards. Meanwhile, the Single Market and the prospect of monetary union have injected new urgency into pressure for economic development at the periphery, while at the same time, demands for environmental policy harmonization have increased.

This situation has produced, not surprisingly, contradictory policy responses. The EU has thus acted as both a catalyst and a resource for environmental improvement in the south. This dual response has created special problems for the southern member-states. These problems arise partly from the fact that the EU's own policy has in a particular sense been contradictory. Its developmental policy, in the form of the Structural Funds, has not harmonized well with its environmental policy. This lack of harmonization may arise partly from institutional fragmentation in the EU decision-making process and partly from the fact that the principle (in the Single European Act) of an environmental dimension to different EU policy areas had not yet been officially adopted when the Integrated Mediterranean Programmes (IMPs) were launched in the mid-1980s. In fact, the regulation establishing them (2088/85) refers to environmental protection as a subsidiary activity to be achieved largely through other sectoral programmes. It did not encourage the submission of proposals which made environmental objectives their principal goal, nor did it specify the general balance between development aims and environmental constraints [*Baldock and Long, 1987: 25*].

The particular method adopted by the EU Commission, of issuing directives, requires their incorporation into national legislation before they become effective, allowing a certain discretion over the form of detailed application. Thus, the EU has a potential for shaping, even modifying, national policy styles, while leaving room for flexibility or possibly national foot-dragging. It is easy to see that the kind of institutional and policy-infrastructural problems, discussed in the previous section, have complicated the process of applying and especially enforcing EU legislation at the national level.

It is against this background that southern European difficulties in adapting to increasing EU environmental legislation become more understandable, although they are not necessarily unique in the European context. A report on the success rate of member-states in transposing EU directives into national laws up to the end of 1991 gave Italy 59 per cent compared with 93 per cent for Spain, 94 per cent for Portugal and 76 per cent for Greece. The figures for selected northern member-states were: Germany 92 per cent, Netherlands 95 per cent, Denmark 98 per cent, and Britain 85 per cent (*The Financial Times*, 3 November 1992). Thus, there was no clear north/south divide on this basis. Italy clearly has the worst record among member-states of applying EU directives on the environment, although recently things have improved with the *legge comunitaria* tightening up parliamentary procedure on EU legislation. As to infringements, Spain has been at the top of the list of member-states against which proceedings have been taken concerning EU environmental laws (*The European*, 26–28 October 1990; *Pais Internacional*, 9 February 1990).

There is a noticeable feeling in the south that EU environmental policy is rather northern in outlook and specifically German in its emphasis on uniform standards. This concern has been expressed over car emission standards and more recently the directive on recycling packaging [*Arp, 1991: 17–18*]. Philosophical differences also surfaced in 1992 over the debate about introducing a carbon tax, arousing southern fears about what this might do to economic growth and unemployment in weak economies struggling to keep pace with the drive towards a single currency (*The Times*, 3 March 1992). There was strong opposition from Spain, concerned about the cost effects particularly on its languishing coal industry (*The Financial Times*, 28 January 1992). Admittedly, though, this was not a straightforward south versus the rest problem, for the proposal – pushed fervently by Commissioner Ripa di Meana – was deeply divisive, such as in the Commission itself over the rival claims of economic growth and environmental rescue (*The Times*, 3 March 1992).

One increasing problem facing the southern countries is the costs incurred in applying EU environmental directives. These costs hit them severely both because they tend to be the poorer member-states, but also they have to make greater strides to keep pace with European environmental policy. It is for this particular reason that Spain blocked the extension of majority voting on the Council of Ministers in this area (*The Economist*, 16 November 1991). This special problem has now become officially recognized in the new Cohesion Fund.

More recently, the EU has begun to develop special programmes for assisting environmental progress, such as MEDSPA, LIFE and ENVIREG, so that the Cohesion Fund follows a certain pattern. EU assistance, such as under the STEP programme, has undoubtedly been important in furthering environmental research in the poorer countries. Furthermore, environmental considerations have begun to be incorporated into the criteria for approval of projects under the Structural Funds. At the same time, there are more regular warnings from Brussels over the environmental impact of construction projects; while Italy has come under more persistent pressure from the Commission for flouting EU drinking water standards on pesticide levels (*The European*, 31 January 1992).

There are two problems that may be observed in southern Europe: governmental incapacity and overload, magnified by policy pressure from Brussels; and some reservations about the pace if not the content of EU environmental policy, particularly from Spain. However, the Union has in effect adopted the carrot and stick approach, and, very gradually, this has started to have limited results.

5. THE SOCIETY DIMENSION: TOWARDS ECOLOGICAL MODERNIZATION?

'Society' as employed here is a collective term for a variety of influences and actors – political, economic and social – which are seen as normally playing a significant part in policy debate and input at the national level. They include political parties as social actors, economic interests, environmental organizations and the media. For most of these actors the EU has in recent years acquired more political salience and impact. The impact of the EU on wider public opinion is more diffuse and more difficult to estimate. It is essentially transmitted via the kind of actors just listed.

We are probably looking at gradual societal response to the demand for greater environmental commitment. It is unlikely, therefore, that the three southern European countries will have moved far along the road of ecological modernization. However, dramatic events can shift opinion, this clearly being the case with the Chernobyl crisis of 1986. For reasons of geographical proximity this had a noticeable effect on Italy, in particular mobilizing elite and informed opinion especially in the northern areas.

How the EU fits into this scenario is obviously complex. But we are not as such concerned with whether the EU has any deeper impact or is out of touch with public opinion – an issue raised in the debate over the Maastricht Treaty – but with how the different actors listed above interact with the EU and among themselves to encourage, or otherwise, policy movement in a more environmentally friendly direction.

The main *political parties* in these countries have tended not to give a priority to environmental policy or concerns, save on an *ad hoc* basis. This is blatantly the case in Spain where 'politicians are convinced green measures do not win votes' (*The Economist*, 27 April 1991). The same is broadly true of Greece (except when one or other party has made a passing issue of air pollution in Athens); while in Italy the parties have not traditionally been any more responsive, although the situation has now begun to shift. That is because of the political arrival of *I Verdi*, the Greens, who have been in the Parliament since 1987 – always a sure indicator of a likely response by political rivals. Growing environmental sensitivity has occurred at a time of widespread political protest and general disaffection with established parties in Italy. Traditionally, political parties were often allied with economic interests and pandered to consumer values. Even in 1990, one study of the Italian environment noted that 'the country's main political forces still regard environmental protection as external, peripheral or only partially relevant to the production-distribution-consumption function of society' [*Alexander, 1991: 106*].

Shifts in the balance of forces in a party system or in the outlook of individual parties may as a rule prove the decisive factor in bringing about change on environmental matters. Apart from that, movement is likely to occur as a result of the following: environmental organizations becoming more effective in lobbying or provoking policy-makers; a gradual sea-change coming from the spread of environmental information or education; adoption by the media of the environment as a regular rather than spasmodic concern; public opinion developing greater environmental awareness in response to these influences or simply environmental events; or a change of attitude on the part of business and industry. Clearly, there is a potential interlinkage between these different factors, but we examine them briefly in turn.

Environmental organizations are fairly numerous in these countries, there being around 700 in Spain according to a 1989 survey [*La Calle Dominguez, et al., 1991*], but their political impact has usually been very limited and they have often appeared as less aggressive than their German and British counterparts (*The European*, 26 March–1 April 1992). Admittedly, the closed decision-making process on the environment, in Spain and Greece particularly, has left only limited scope for lobbying. The passivity of these organizations has otherwise been illustrated by the link between private complaints about environmental problems and activism on the part of such organizations; that is, the latter perform a mobilizing function with respect to the former. Complaints are noticeably infrequent in the south which may also owe something to basic public scepticism towards or mistrust of state authorities. However, an opportunity has been opened up at the European level, for the EU Commission has come to rely considerably on environmental organizations for concrete information on environmental defaults which is then used against member-governments. Those in southern Europe have also begun to respond to this channel for influence.

As to *public information*, the environmental organizations clearly have a part to play as do governmental authorities. The relevant ministries in Rome and Madrid, while issuing reports and publicity on environmental matters for some years, have also invested in environmental education (the Greek government has been rather less active). More recently, environmental education in schools has developed in Italy and to some extent in Spain. The ministry in Rome has promoted this as well as vocational training in environmental matters [*Ministero dell'Ambiente, 1989: Part III*]. It is difficult, however, to establish whether any gradual sea-change is occurring, as the effects of these relatively new activities are likely to be slow. Some clues may come from the role of the media and changing patterns of public behaviour.

The *media* have their own sensationalist way of interpreting issues, as in Italy, where headline news has concentrated on major emergencies and crises such as the problem of the algae in the Adriatic or the affair of the 'Karin B' waste disposal ship. The press has nevertheless served an important purpose in persistent coverage of scandals, such as the death of thousands of wild birds in the Donana nature reserve in south-west Spain. Since the mid-1980s there has been a distinct growth in regular press coverage in Italy of environmental matters outside emergencies; 'urban smog' has been a regular for several years, and informative background reports ('dossiers') have occasionally featured, for example on drinking and coastal water. By comparison, the national press in Spain has been less assiduous in its coverage. Since one of the problems there has been access to hard information, Spanish journalists have come to rely more on environmental organizations than on government sources [*Aguilar, 1992: 18–19*]. As a whole, television has usually been much more neglectful of environmental matters than the quality press. The 1992 Ministry report on the state of the environment in Italy quoted a survey on the state radio and television network (RAI) during 1986–90, showing that environmental problems counted for only 1.7 per cent of news items [*Ministero dell'Ambiente, 1992: 26*]. There is of course an environmentalist press in these countries, but that tends to reach only the converted.

One feature remarkable in southern Europe is the strong *localistic focus* on environmental problems, suggesting a link between territory and environmental awareness. This is noticeable in Spain in that, in several areas, the regional press is often more interested in such problems than is the national press. The Italian ministry carried out a survey on the selected local press during 1989–91 [*Ministero dell'Ambiente, 1992: 450–1*]. It emerged that waste disposal was the most itemized issue, clearly one of particularly local concern. In 1989, there was, for instance, a major local scandal at Montalcino, Tuscany, where the proposal to establish a large waste disposal facility aroused intense local opposition, concerned that it would contaminate the soil in this famous Brunello wine-growing area. The localism in southern Europe was also evident in the campaigns conducted for the 1987 European Year of the Environment. In Spain, they included ambitious regional programmes financed by several autonomous communities, especially on nature protection; in Italy, there were numerous local projects (monuments of Rome, green areas in Naples, pollution in La Spezia); while in Greece, the 'campaign in general was far more local and national rather than European', including much activity by small groups and NGOs at local level.

Undoubtedly, the various actors discussed above have an influential role in sensitizing people to the environment, but whether this has any deeper impact is less clear. The signs from southern Europe are mixed, with some evidence of growing environmental consciousness, but hardly – as yet – any overall 'remaking' of public attitudes over the environment, save possibly in Italy. Even there, consumerist attitudes have dominated, as the familiar attachment to the automobile in Italian life testifies: 'The car is a consumer good surrounded by a mythical halo, as an imaginary collective that has a consistent part in influencing the level of demand . . . the point of attack on the culture of the car is precisely its role as a status symbol (and bad habit) . . .' [*Lega per l'Ambiente, 1990: 30*].

From the late 1980s, there were signs of some change with a growth in Italy of the use of lead-free petrol (though not in Spain and Portugal), the appearance of bottle banks in rural towns as elsewhere and the institution in some Italian cities of traffic-free zones (*Die Zeit*, 17 June 1988). Local authorities played an important part in such changes, and they seemed to form part of some attitudinal transformation. One report in 1989 noted that 'nowhere in Europe is the environmental pendulum swinging so fast, from neglect to acute concern, as in Italy' (*The Financial Times*, 24 April 1989). But this was regionally variable, for 'environmentalism is strongest in the North of Italy and weakest in the south' [*Alexander, 1991: 105*], which may be explained in terms of greater proximity to central and northern Europe, where environmental values are generally more developed. Familiar socio-economic differences between the north and south of Italy are also likely to have been influential here.

Environmental awareness is rather less advanced in Spain than in Italy for several reasons: a lack of environmental education in schools, the overwhelming emphasis on economic growth, and the lack of participatory mechanisms combined with traditionally hierarchical relations between public administration and society [*Aguilar, 1992: 22–4*]. Others have noted that Spaniards see their country as largely empty, so that dumping rubbish in the countryside is hardly regarded as offensive (*The European*, 28–31 May 1992). However, opinion polls have in the last few years detected some signs of an increase in ecological sensitivity. A *Demoscopia* survey in late 1990 identified a growing awareness but, at the same time, a firm reluctance to accept more taxes on cars and petrol (*Pais Internacional*, 8 October 1990). In Greece, on the other hand, the 1980s have seen a gradual rise in awareness from the time the OECD report noted that 'public concern about environmental problems is fairly recent in Greece but is an increasingly important force' [*OECD, 1983: 12*].

Turning to *business and industry*, the Commission's Task Force report of 1989 noted the low level of environmental markets in southern Europe

[*EC Commission, 1989: 9.3*]. Clearly, business and industry are more likely to adapt when both the global markets and national publics demand more environmentally friendly products. In Italy, on other counts the most environmentally advanced of the three countries, the evidence so far has not been encouraging. One recent survey of four European countries recorded that the Italian market was far less ready to follow tighter norms over car emissions than the German, and that this affected the attitude of companies [*Arp, 1991: 13*]. A 1990 report on environmental policy in relation to socio-economic planning noted a conflict between growing environmental consciousness and industrial secrecy in Italy, especially over soil protection [*Lega per l'Ambiente, 1990: 510*]. Such conflict seems to be under increasing pressure of EU legislation. Changing attitudes is not the only problem. There is also the question of capacity for environmental adaptation. Much of industry in these countries has neither the skills nor resources to meet changing international demands, particularly the high proportion of small and medium businesses with very limited investment scope and weak technology [*Estevan, 1991: 475*]. Until recently, Spanish industry 'considered the protection of the environment as an extra cost, without any productive yield'; however, attitudes have begun to change and the environment variable is being viewed as a means for introducing improvements in products [*Estevan, 1991: 476*]. FIAT, in Italy, is exceptional in having responded more clearly since the mid-1980s to European markets. Greece has a less complicated experience since it is less industrialized and has no car industry. Also, an initiative was taken in the mid-1980s to regulate its shipping industry to protect the marine environment of the Mediterranean [*Rehbinder and Stewart, 1985: 212*].

Overall, then, the southern European countries have, to differing degrees, evidenced greater mass-level interest and activity by interest and pressure groups. But this is a recent departure. It is one that may well continue, and if so could lead to a change of policy direction, although the impact of the recession has to caution any judgement about the immediate future.

6. CONCLUSION

Discussion of the three southern member-states of Italy, Spain and Greece has certainly demonstrated the complexities of EU policy-making as emphasized by the quotation introducing this chapter. In particular, the EU's policy process is crucially dependent on national institutions and procedures – which in these countries have serious weaknesses – and these may have a profound effect on policy outcomes in the EU.

We have hypothesized that national governments are more likely to lend the environment a priority and to change their policy approach when there is combined pressure from outside (the EU) and from below (society). In the three cases examined, these pressures are real and indeed stronger than ever before; but they are also limited and vary cross-nationally.

Policy in the southern member-states has tended to be reactive towards EU policy demands on the environment, perceived by them like a cold if bracing wind from the north. However, it is undeniable that the EU is responsible for a considerable amount of the environmental legislation in these countries as well as for modest moves towards environmental planning. The EU has thus been significant in the content of, and to some extent the motivation behind, environmental policies in the south. The record is less clear as to the priority and coherence of their policies, for EU membership has rather acerbated conflict between economic concerns and environmental ones. The southern importance attached to the economic imperative has remained, although there are recent signs of policy innovation, especially in Italy and in some regions there and in Spain. These have combined with signs of greater public, media and even interest and pressure group sensitivity to environmental concerns.

The idea of there being a 'north/south divide' between member-states over the environment can be overrated. The southern countries do have particular problems of administrative procedure and competence and they are notably short of infrastructural facilities in this policy area. They also face difficulties in meeting environmental costs. But, in other respects, they stand less apart from the northern states. This is certainly true of the problem of fragmented policy structures, whether at national or centre-periphery level. Furthermore, environmental policy priority may vary among member-states irrespective of such a 'divide' (it has conceivably been stronger in Italy than the UK in the past half-decade).

Moreover, there are also limits to regarding the southern three as simply one group. There are some significant differences among them, over not only policy priority, but also policy structures, enactment of EU legislation and environmental awareness, not to mention policy content, given differing environmental concerns. If anything, recent changes in the environmental field have tended to highlight such differences. Apart from policy innovation, these are most evident at the societal level, with Italy again more advanced than the other two. It is difficult, however, to attribute all these changes simply to that country's longer membership of the EU. The impression gained from this study is that deeper change will come about more as a result of general secular processes – such as European or other influences of a vaguer transnational kind – than of policy initiative from Brussels. That reminds us not to be too deterministic about the direct effects of EU policy on member-states.

ACKNOWLEDGEMENTS

This chapter is part of a research project, Environmental Standards and the Politics of Expertise in Europe, funded under the Single Market Programme of the Economic and Social Research Council (ESRC). The author wishes to thank Albert Weale for his comments on a draft version.

REFERENCES

Aguilar, S., 1991, 'Policy Styles and Policy Sector Influence in Pollution Control Policies', paper presented at the ECPR Joint Session of Workshops, Essex, 1991.
Aguilar, S., 1992, 'Environmental Monitoring and Environmental Information in Spain', in H. Weidner, R. Zieschank and P. Knoepfel (eds.), *Umwelt-Information* (Berlin: Sigma).
Alexander, D., 1991, 'Pollution, Policies and Politics: The Italian Environment', in F. Sabetti and R. Catanzaro (eds.), *Italian Politics: A Review*, Vol.5 (London: Pinter).
Arp. H., 1991, 'Interest Groups in EC Legislation: The Case of Car Emission Standards', paper presented at the ECPR Joint Session of Workshops, Essex, 1991.
Baldock, D. and Long, T., 1987, *The Mediterranean Environment Under Pressure: The Influence of the CAP on Spain and Portugal and the 'IMPs' in France, Greece and Italy* (London: Institute for European Environmental Policy).
Bennett, G., 1991 (ed.), *Air Pollution Control in the European Community* (London: Graham and Trotman).
Capria, A., 1991, 'Formulation and Implementation of Environmental Policy in Italy', paper presented at the 12th International Congress on Social Policy, Paris, 8–12 October 1991.
del Carmen, M., 1986, 'Spain's Accession to the European Community: Repercussions on Spanish Environmental Policy', *European Environment Review*, Vol.1, No.1, October, pp.13–17.
La Calle Dominguez, J., et al., 1991, *On The Origins of the Environmental Question in Spain* (Madrid: October).
EC Commission, 1989, *Task Force Report on the Environment and the Single Market*, Luxembourg.
EC Commission, DG XI, 1993, *Administrative Structures for Environmental Management in the European Community*, Luxembourg.
Elsom, D., 1987, *Atmospheric Pollution* (Oxford: Basil Blackwell).
Estevan, M., 1991, *Implicaciones Economicas de la Proteccion Ambiental de la CEE: Repercusiones en Espana* (Madrid: Ministerio de Economia y Hacienda).
Hine, D., 1993, *Governing Italy: The Politics of Bargained Pluralism* (Oxford: Clarendon Press).
Lega per l'Ambiente, 1990, *Ambiente Italia 1990*, Giovanna Melandri (ed.), (Milan: Arnoaldo Mondadori).
Lewanski, R., 1993, 'Environmental Policy in Italy: From the Regions to the EEC, a Multiple Tier Policy Game', paper presented at workshop 'Environmental Policy and Peripheral Regions of the EC', ECPR Joint Session of Workshops, Leiden, 1993.
Lopez Bustos, F., 1992, *La Organizacion Administrativa del Medio Ambiente* (Granada: Editorial Civitas).
Ministerio de Obras Publicas, 1989a, *Medio Ambiente en Espana, 1988*, Madrid.

Ministerio de Obras Publicas, 1989b, *El Derecho Ambiental y sus Principios Rectores*, Madrid.
Ministerio de Obras Publicas, 1990, *Medio Ambiente en Espana, 1989*, Madrid.
Ministero dell'Ambiente, 1989, *Rapporto al Ministro sulle Linee di Politica Ambientale a Medio e Lungo Termine*, Rome.
Ministero dell'Ambiente, 1989a, *Rapporto sullo Stato dell'Ambiente*, Rome.
Ministero dell'Ambiente, 1992, *Relazione sullo Stato dell'Ambiente*, Rome.
Ministero dell'Ambiente, 1992a, *Bilancio di un Quinquennio di Politiche Ambientali*, Rome.
Minister of Environment, Athens, 1991 *National Report of Greece*.
Nanetti, R., 1990, 'Social, Planning and Environmental Policies in a Post-Industrial Society', R. Leonardi and R. Nanetti (eds.), *The Regions and European Integration: the Case of Emilia-Romagna* (London: Pinter).
OECD, 1983, *Environmental Policies in Greece*, Paris.
Rehbinder, E. and R. Stewart, 1985, *Environmental Protection Policy*, Vol.2 (Berlin: Walter de Gruyter).
Wallace, H., 1977, 'National Bulls in the Community China Shop', in H. Wallace, C. Webb and W. Wallace (eds.), *Policy-Making in the European Communities* (Chichester: John Wiley).
Weale, A., 1991, 'Ecological Modernization and the Integration of European Environmental Policy', paper presented at conference on 'European Integration and Environmental Policy' at Woudschoten, The Netherlands, November.
Williams, A., (ed.), 1984, *Southern Europe Transformed* (London: Harper and Row).

Spanish Pollution Control Policy and the Challenge of the European Union

SUSANA AGUILAR-FERNÁNDEZ

INTRODUCTION

In the last two decades Spain has undergone a very significant transformation of an essentially political nature. The return to democracy has entailed the modernization of an archaic political structure, alien to a society which experienced a profound economic and social change in the 1960s. Within this political transformation, two facts, one at the domestic level and the other at the international level, deserve special attention:

1. The creation of the *estado de las autonomías* (literally, 'state of autonomies'), and the subsequent setting up of 17 regional governments (the so-called *Comunidades Autónomas* CAs); this process began after the approval of the constitution in 1978, and pursued two goals: the satisfaction of specific historical demands for political autonomy, and the achievement of a general situation of decentralization.
2. Entry into the European Community (EC, now European Union or EU); membership of the Community meant the achievement of one of the most enduring aspirations in Spanish politics.

These two facts have radically influenced the political scenario and can be traced, for instance, in pollution control policy.[1] In this policy the process of decentralization has led to the establishment of new environmental administrations at the regional level. Regions have become the main units responsible for the implementation of a policy whose main source, since 1986, is Brussels. The entry into a supranational organization which has greatly contributed to the upgrading of environmental issues, has prevented the Spanish government from neglecting pollution control policy any longer. This policy entails an important cost because Spain is not only pressed (as are all EU member-states) to clean up and restore its damaged natural assets, but also, and more specifically, to preserve its unique natural habitats, even at the price of leaving the economic development of those areas aside. However, this cost will be accompanied by EU aid, as was decided at the Maastricht Summit, with the approval of the Cohesion Fund.

The Cohesion Fund has been devised to meet the needs of the less advanced countries of the EU in the fields of public works and infrastructure, and environmental protection. In these countries, poorer economic conditions coincide with a geographical situation far removed from the centre of the EU. These factors have normally been used to define peripheral countries. Accordingly, Spain would be peripheral in two senses:

1. The country still has to bridge the economic gap in relation to the countries at the core of the EU.
2. The country is not only relatively distant, geographically speaking, from the centre of decisions in the EU, but it is also insufficiently equipped to defend its stance on some Union issues like pollution abatement.

The inadequacy of means to influence EU policy-making in certain areas is accompanied by an inadequacy in the implementation of EU policies. This implementation deficit is particularly evident in pollution control policy.

APPROACH TO A CHARACTERIZATION OF POLLUTION CONTROL POLICY: ITS PECULIARITIES IN SPAIN

Pollution control policy is a relatively new area that has acquired great relevance in a short time. This policy can be characterized as a complex area in which experts enjoy a significant role because of the importance of expertise as a political resource. Likewise, complexity also accounts for the privileged role of industry in the political scenario; industrialists, unlike most environmental officials, have the required technical knowledge to make and implement this policy and are also strongly associated with experts. This explains why industry[2] has been one of the main actors in a policy that has generally developed in closed and somewhat secret meetings enjoying, consequently, a rather peaceful atmosphere. This peace has been contributed to by the general lack of influence exerted by other 'more confrontational' actors, such as the environmental groups.

Another feature of pollution control policy is the far-reaching effect of the EU, which strongly influences the development of member-states' environmental policies. This influence is evident in, among other things, the growing difficulties that nations face in formulating these policies autonomously.

The general characterization of pollution control policy as a complex area that is basically in the hands of industry (and, of course, government) and subject to international influence, applies to most EU member-states. However, it has to be specified in relation to the different countries. In the Spanish case, for instance, this characterization is expressed in a slightly different way. In Spain, industry has more weight than other private actors but, unlike in other European countries, it does not enjoy institutionalized participation in the pollution control policy process. More precisely, government agencies do not usually rely on business associations or private sector companies to elaborate and enact environmental regulation. The pre-eminence of public authorities has been to the detriment of private interest groups, particularly environmentalists. Although some attempts have been made to integrate environmentalists into the policy-making process, this sector has traditionally had very weak influence in the pollution-abatement field.

The lesser role of industry in the environmental policy process, comparatively speaking, goes along with a general lack of co-operation between public and private actors in Spain. This pattern, which will herein be called 'statist institutional design', resembles the model of social order described by Streeck and Schmitter [*1985*]. This model is identified by a central institution (state or bureaucracy) which embodies a guiding principle (hierarchical control). The explanation of this design in Spain has to take two basic factors into account: the historical tradition of state intervention in the public sphere, and the long-standing irrelevance of interest groups in the political arena [*Linz, 1981*].

Spain also shows some peculiarities with regard to the control exerted by the EU on the pollution control policy of its member-states. The EU's influence on this policy is resisted more strongly, though not necessarily more successfully, in Spain than elsewhere. Spain was the last state to join the EC (in 1986) and, consequently, had not contributed to the development of a policy, the beginning of which can be dated to 1972. In addition, environmental policy was traditionally considered a secondary issue at the domestic level. But it is not only the newcomer status of Spain that explains its poor contribution to the upgrading of pollution control policy. Indeed, the country has no interest in promoting a policy which could, allegedly, impinge harmfully upon its economic situation. Spain's peripheral condition has fostered the consideration of environmental policy in terms of a trade-off. This trade-off would assume an uneasy compatibility between economic development and nature protection, and would also aim at softening the second element so as not to disturb the first.

The fact that some member-states consider environmental policy in terms of a trade-off has not weakened a policy which s highly regarded by most European citizens. Besides, Spain has not managed to modify environmental priorities because its lack of significant political power (or negotiating force) within the EU framework has made it an excellent recipient of other member-states' initiatives in this field. For this reason, the Spanish government tries to oppose, or postpone, the influence of a union policy that is mainly perceived as the answer to other countries' worries. The environmental problems of the peripheral countries, which basically embrace soil erosion, desertification, and forest fires,[3] have been traditionally put aside in comparison with the problems of the countries at the centre; namely, air pollution, waste management, control of chemical substances, and so on. Only recently have the problems of the periphery acquired a relative importance (as is shown by the MEDSPA fund, and the final declaration of the Dublin Summit in 1990) and reached a priority status [*Kraemer, 1988*].

THE STATIST DESIGN IN POLLUTION CONTROL POLICY

Administrative Disorganization in Pollution Control Policy

The environmental administrative structure is mainly characterized by a dispersion of responsibilities in diverse agencies of different central government ministries. Measures introduced to overcome this dispersion during the 1970s were unsuccessful, and the same problem reappeared later, in different ways, in the administrations created by the regional governments (CAs). Later on, once the new regional structure was set up and Spain became a member of the EC, the first attempts to co-ordinate the different environmental organizations took place, and the administrative structure at the central level was upgraded.

The administration for the protection of the environment has traditionally been composed of heterogeneous public agencies, that could only partially be classed together as pursuing the same goal [*CIMA, 1978*]. The predominant feature of this administration was a lack of co-ordination both between and within sectors. In an attempt to solve this situation, the Interministerial Committee for the Improvement of the Environment (CIAMA) was set up in 1971, but one year later was replaced by the Interministerial Commission for the Environment (CIMA). Although this commission represented the first real effort aimed at co-ordinating a policy distributed among a multitude of ministries with divergent goals, its efficiency was quite low. A number of reasons (overload of tasks, no compulsion to produce reports, lack of funds, and so on) kept this organization from functioning as a co-ordinating body until its disappearance in 1987. During this period, along with the languishing development of the CIMA, some changes occurred at the central level. In 1977 responsibility for environment was transferred from the Presidency of Government to the Ministry of Public Works (MOPU) and, one year later, the Head Office for the Environment (DGMA) was created.

The creation of the regional environmental administration began in the mid-1980s and it responded to the need to accomplish a constitutional dictate. According to the constitution, the regions in Spain can assume a wide range of environmental responsibilities: they are also responsible for environmental management and are able to set additional regulations in this field. The central government, on the other hand, has exclusive responsibility for basic legislation, policy co-ordination, and international representation concerning environmental policy. In the course of the creation of the new administrative structures, the regions have chosen different models. However, most regional governments have consolidated the dispersion of environmental responsibilities that already existed at the central level. Along with this dispersion, the situation has become more confused due to the lack of vertical co-ordination prevailing between the central and the regional levels.

The entry of Spain into the EU has fostered the putting into practice of several projects that aimed at overcoming the traditional deficient co-ordination, both horizontally (among different ministries with environmental responsibilities in the central government), and on the new vertical axis (between central government and the regions). The projects for providing horizontal co-ordination no longer exist, due to the disappearance of the CIMA, and the inefficiency shown by various sectoral co-ordinating commissions [*Martínez Salcedo, 1989*]. Nonetheless, this deficiency in co-ordination has been somehow alleviated by the recent upgrading of the environmental administration at the central level, and the consequent concentration of tasks in one single government agency. In 1990, the General Department of the Environment (SGMA) was set up, assuming the former responsibilities of the DGMA. One year later, as a result of the transformation of the MOPU into the MOPT (Ministry of Public Works and Transport), the State Department for Water Policies and the Environment replaced the SGMA. Finally, after the last general elections held in June 1993, this unit was renamed as the State Department for Environment and Housing.

The projects for providing vertical co-ordination began at the end of 1986 by means of some informal meetings where central and regional environmental officials gathered. These meetings were organized by the MOPU, and took place at two different levels: among technicians, mainly as a framework for the exchange of information; and among general directors, for global policy co-ordination. In 1987, the main public agencies responsible for this area gathered at a Sectoral Conference of the Environment which has been convened several times since then. Yet only in 1991 did this conference produce a document that put forward, for the first time, the shared objectives of environmental policy in Spain [*Mopu Informa, 1991*].

The recurrence of the attempts to overcome the lack of co-ordination (be it horizontal or vertical) can basically be attributed to the EU. The European Commission has often referred to deficient co-ordination as one of the main causes of the environmental policy implementation deficit in Spain. An additional cause is the already mentioned statist design, which also prevails at the regional level to varying degrees. These two factors – statist design and inefficient organization – help to explain why the implementation deficit, that pervades environmental policy everywhere, is more acute in Spain than in other EU member-states.

Unilateral Policy-making and Negotiation in the Implementation of Pollution Control Policy

In Spain, the statist design in pollution control is traceable in a policy-making process in which government agencies play a predominant role, and in which no institutionalized relationship, or formal co-operation, has been established between public authorities and industrial associations. Nor have groups of experts (outside the administration) been regularly consulted; the participation of specialists in the policy process is not a customary practice of the administration [*Martín Rebollo, 1984*].

At the central level, there have been several projects that provided for the creation of environmental agencies allowing social participation. Yet, most projects were not passed in the end, and the few organizations created functioned unsatisfactorily until their final suppression. The General Law of the Environment, which embraced the constitution of a national commission on the environment, was finally discarded [*Costa, 1985*]. The Project of Basic Criteria for the Protection of the Environment, which posed the creation of a high council for the environment, did not come into force either [*Boletin Informativo del Medio Ambiente No.14, 1980*]. The only public body of plural composition which was finally set up, the Committee of Public Participation, was suppressed in 1986, three years after its foundation, and did not play any significant role in this policy [*Información Ambiental No.9, 1986*]. More recently, however, in October 1993, the government announced the creation of an advisory body on environmental issues which will allow environmentalist participation.

At the regional level, the situation relating to social participation in environmental policy is similar [*Aguilar-Fernández, 1993*]. Most of the councils for the protection of the environment have representatives of the administration, but do not include private groups, be they industrialists or environmentalists [*Suárez, 1990*]. Basically, the only institutions of mixed composition are the regional councils of the environment. These councils, generally integrated by environmentalists and citizens' associations, limit themselves to conservationist tasks. There are, nonetheless, several projects that provide for the creation of similar agencies whose tasks would not be limited to the protection of natural resources (for example, the Social Council of the Environment in the Basque Country). Existing organizations of this kind, such as the Advisory Council of the Environment in Andalusia, have nonetheless had quite an unsatisfactory development.

To summarize, in pollution control policy there is no permanent and specific administrative unit that allows industrial participation, nor is there a commission of mixed composition for the discussion of environmental policy. The relationship between government and industry is informal and discontinuous, and usually unfolds, not through associations, but through personal contacts. These contacts normally aim at solving urgent and specific issues. They have not, in most cases, resulted in the creation of informal networks which could somehow substitute for the lack of formal structures of participation.

The absence of an institutionalized design based on the participation of industry does not necessarily mean that the administration is a monolithic block. The government exhibits inter-agency conflicts due to the lack of agreement on the weight that should be attributed to industrial associations in pollution control policy-making. This disagreement is most salient if two agencies at the central level are taken into account: the MOPT, the main organization responsible for environmental protection, and the Ministry of Industry (MINER), with sectoral tasks in this area. Between these two ministries there is a sort of division of labour. The former is basically connected with the Chambers of Commerce, and considers one of its main tasks to be to reach an environmental pact among the wide array of involved social interests [Aguilar, 1992]. The latter is more sympathetic to the industrial world and has a better relationship with the Spanish Confederation of Business Associations (CEOE). Yet the fact that different public agencies show different attitudes towards private industrial groups does not substantially alter the pattern, described above, of a lack of standing co-operation between public and private actors.

The statist institutional design, predominant at the policy-making stage, is also reflected in the implementation of pollution control policy. Two basic facts reveal the presence of statism:

1. Government reluctance towards industrial self-regulation;
2. The scarce application of voluntary measures (understood as environmental objectives that, once posed by the government, can be voluntarily assumed by industry and pursued with its own means and procedures).

In the implementation process, however, and despite the exclusion of industry from the policy-making process, there has been relative cooperation – not exempt from a certain degree of conflict – between regional officials and specific firms. This co-operation can mainly be explained by two reasons:

1. The low political priority of this area, which has made governmental and industrial interests basically coincide (that is, neither public nor private actors have felt compelled to enforce or to abide by environmental regulation, respectively);
2. The features of the policy area, which foster mutual dependence between government and industry (that is, public actors need to know the environmental situation of the firm, whereas private actors are not able to keep track of the changing flow of environmental rules, and are also dependent on public finance for the adoption of anti-pollution measures).

This situation favours the recognition of a common interest which has to be jointly pursued. The need to accomplish this common task has led to a process of informal negotiation between experts from both administration and industry, as a step prior to implementation. Some results of this negotiation process, which is kept away from political debate and public scrutiny, are the approval of gradual plans for firms not abiding by the law to adapt to environmental legislation, the granting of public subventions to industry for the adoption of pollution abatement measures, the prior notification of monitoring visits, and the rare application of sanctions [*Aguilar-Fernández, 1993; Mayntz, 1978*].

In general terms, the measures taken in the process of negotiation can be distinguished according to two criteria:

1. In scope, the measures are either sectoral or global; sectoral if they are limited to specific steps for pollution abatement, and global if they affect the whole industrial process (from cradle to grave).
2. In nature, the measures are either compulsory or voluntary; compulsory if their lack of fulfilment leads to sanctions, and voluntary if they are based on compromises not entailing any legal obligation. The combination of these two criteria is shown in Figure 1.

In Spain, most environmental measures have a sectoral and compulsory character. There also exist global compulsory measures (environmental pacts), whereas voluntary ones, whether sectoral or, in particular, global, are rather exceptional. These measures can be classified as follows:

FIGURE 1
CLASSIFICATION OF POLLUTION ABATEMENT MEASURES

	Nature:	
Scope:	Compulsory	Voluntary
Sectoral	Environmental agreement	Voluntary agreement
Global	Environmental pact	Voluntary pact

1. Environmental agreements: pollution abatement measures that are carried out, with public aid, between a CA or a municipal council and a firm (a two-tier agreement), or between these types of agency and the central government (a three-tier agreement).
2. Environmental compromises, which could be considered a subtype of environmental agreements: measures for the construction of waste disposal facilities that are carried out between a CA and different firms of the same industrial sector.
3. Voluntary agreements: agreements between the central government and the firms of an industrial sector for the reduction, or substitution, of toxic substances or environmentally harmful products.[4]
4. Environmental pacts: pacts between a CA and a big firm, or several firms located in the same area, for the adoption of global pollution abatement plans with the main aim of restoring environmentally damaged zones.

THE EUROPEAN UNION CHALLENGE: PRESSURES UPON THE STATIST DESIGN IN POLLUTION CONTROL POLICY

Traditionally, the government in Spain had formulated and carried out pollution control policy without taking industrial interest groups significantly into account. Industry, on the other hand, had not made sufficient effort to get involved in the political process because of the general low priority attached to the issue. Entrepreneurs were also fearful about the compromises that an institutionalized relationship with the government could entail. However, the relevance of the EU as 'the only environmental policy-making institution in the world with the power to impose binding obligations on sovereign nation states' [*Bennet,* et al., *1989: 11*], has begun to make itself felt in Spanish environmental policy. In this sense, the entry of Spain into the EU is putting pressure on an institutional design in which public agencies had an almost exclusive role in the public sphere. The EU is forcing the government to adopt a more stringent policy, so that the environmental gap with other member-states should diminish. Likewise, the EU has caused industry gradually to recognize the importance of environmental issues.

As a result of the relevance attached to environmental issues in the EU scenario, some concern is growing in Spain. The concern stems mainly from the fact that the country belongs to the group of member-states with the highest degree of non-fulfilment (enactment plus implementation) of environmental laws. Although the system for controlling the application of EU environmental legislation is far from satisfactory, it constitutes an important indication of the degree of law fulfilment by the member-states. For this reason, attention should be paid to the fact that Spain had, in 1990, the highest number of breaches of green rules (57) and, in May 1991, was still the country that most violated this legislation (66 breaches: see Table 1).

TABLE 1
PROCEEDINGS TAKEN AGAINST MEMBER STATES FOR BREACHES OF GREEN RULES, 1990–91

	Warning letters		Reasoned opinions		Court cases		Total		Total difference
	1990	1991	1990	1991	1990	1991	1990	1991	1990-91
Germany	13	1	8	18	8	11	29	30	+1
Spain	45	30	9	25	3	11	57	66	+9
Belgium	27	3	8	17	11	1	46	30	-16
Denmark	5	2	-	2	-	-	5	4	-1
France	28	9	6	13	7	6	41	28	-13
Greece	37	13	5	31	3	6	45	50	+5
Ireland	16	12	5	13	-	3	21	28	+7
Italy	17	16	16	28	7	9	40	53	+13
Luxembourg	9	1	2	11	1	2	12	14	+2
Holland	18	5	5	15	2	2	25	22	-2
Portugal	10	8	4	15	-	1	14	24	+10
UK	18	12	8	9	5	2	31	23	-8
Total	242	112	76	197	44	63	362	372	+10

Sources: Commission of the European Communities, 1990; Sbragia, 1992.

Dynamics of Change in the Statist Design: the Environmental Pacts

The significant environmental deficit in Spain explains why the statist institutional design, despite its long-standing character, has undergone some significant changes. EU membership has led to political reflection on the advantages that the bringing together of public and private actors could entail for the fulfilment of environmental requirements. This reflection is evident in the decrease of government's reluctance towards social participation in this policy area, as well as in the higher receptivity of industry to this likely participation. More precisely, the EU has facilitated the emergence of two new phenomena: the relative rapprochement between government and industry, and the emergence of environmental pacts.

The bringing together of public and private actors is present in the setting up of contacts between the administration (central and regional) and different industrial associations to deal with environmental issues. At the central level, for instance, meetings between the government and large firms for the formulation of environmental adaptation plans – within the framework of the MINER Programme for the Creation of an Industrial, Technological and Environmental Base – have taken place. Likewise, at the regional level, the first formal meeting between the Basque government and the Basque Business Confederation (CONFEBASK) was held in 1988. More recently, a global agreement for co-operation between the Environmental Agency of Andalusia and the association of industries of this region has also been signed. However, the most outstanding phenomena in this new scenario are environmental pacts.

Environmental pacts have mainly been carried out in the Basque Country, and are becoming established as a new practice in pollution control policy. These pacts, which contain binding and global measures (embracing air and water pollution, and waste management), are carried out in heavily polluted areas and usually affect large polluting industries. They have several actual or potential consequences:

1. They usually promote the adjustment of public and private action (that is, speeding up the connection between government cleansing plans and industrial anti-pollution measures).
2. They tend to boost co-operation among firms for the resolution of common problems.
3. They aim at achieving the quickest and best adaptation of industry to EU requirements, and can also incorporate a preventive approach.

The appearance of environmental pacts has produced a qualitative change not only in the practice of pollution control policy, generally focused on specific and sectoral measures, but also in its institutional design. Pacts usually entail the establishment of regular dialogue, more or less institutionalized, between government and industry. In this sense, these pacts could be seen as a sign of the transformation of the traditional relationship between public and private actors.

THE RESILIENCE OF THE STATIST DESIGN IN POLLUTION CONTROL POLICY

In spite of the fact that environmental pacts have tended to spread and become established, it is not possible to conclude that they will remain as a basic and permanent practice in pollution control policy. The statist design, although subject to pressure, is not likely to change dramatically in the near future. State intervention and the reluctance towards private government persist in different ways and are revealed, for instance, by two facts:

1. The exclusively public composition of the bodies in charge of implementation of the national plan of industrial waste;
2. The attempt of the administration, in several cases, to monopolize the management of industrial residues to the detriment of private initiative.

The resilience of institutional designs can partly be explained by their structural character: they are the outcome of historical factors that are deeply rooted in the political process. Another more specific reason for this pervasiveness is that the EU has, up to now, attached only minor importance to the match between the policy implementation processes and the institutional designs of member-states. Besides, nation-states continue to be the most important actors in the EU scenario, and have managed successfully to retain important political fields under their control.[5] Along with the disproportionate influence of member-states in the EU scenario, the choice of the directive as the main legal instrument of environmental protection also accounts for the resilience of national designs.

Directives presuppose, through their discretional character, that all member-states have a suitable political-administrative organization for the implementation of EU objectives. This presumption is clearly false. In general terms, southern European countries have inadequate administrative designs when facing the need to fulfil environmental policy; they are characterized by a tradition of separation between public and private actors, and weak interest groups. Northern European countries, on the contrary, usually have administrative designs that promote co-operation between public and private actors, and achieve better records in the implementation process. Yet co-operative or corporatist models by no means constitute magical recipes for improving the results of the policy process; they can be prone to agency capture and can easily restrict political accountability to a disproportionately small circle of 'privileged' interest groups. Despite this possibility, and although the application of environmental law is far from optimal across the EU member-states, the records of 'corporatist' countries are, in comparative terms, reasonably good. Holland, Denmark, and Germany, have favourable results in relation to the ease and effectiveness of policy implementation [*Commission of the European Communities, 1990*], whereas Spain, Greece, and Italy are the countries with the highest number of breaches of green rules.

Last, but not least, institutional designs have pervaded, not only because countries play the most important role in the EU scenario and directives leave political implementation processes untouched at the national level. Insufficient funds and personnel in the European Commission are also factors which lead to a deficient control over those processes. In this sense, DG XI, the Head Office in charge of environmental protection, is one of the smallest and more insignificant administrative units within the Commission. Its budget is less than one per cent of the EU budget and it only employs approximately 150 officials [*Bennet et al., 1989*]. In addition, the Enforcement Section, the department for the surveillance of legislation fulfilment, can count on just seven or eight people [*Aguilar-Fernández, 1993*].

THE ROLE OF SPAIN IN THE EUROPEAN UNION SCENARIO: APPROACH TO FUTURE TRENDS

The undeniable influence of the EU in pollution control policies, and the clear economic consequences of the environmental issue, have turned EU environmental policy into a conflict scenario. In this scenario, north European states have played an essential role, to the detriment of southern states. The more powerful countries, with advanced policies and environmentally sensitized societies, have often managed to impose their costly pollution abatement measures upon countries with different problems. One of the most outstanding examples of this success is EU air pollution policy. The directive on emissions from large combustion plants (88/609/EEC) was adopted under German pressure after that country had passed a similar regulation at the domestic level [*Liberatore, 1989*]. The divergence between the centre (the so-called 'troika', Germany, Denmark and Holland) and the periphery (basically, Spain, Portugal, Greece, and Ireland)[6] is consequently very acute in environmental policy. The fact that the core countries have succeeded in transforming their worries into EU priorities helps to explain Spanish resentment against EU environmental policy, and also explains the specific strategy developed by this member-state on environmental issues.

In order to balance this 'biased' environmental policy, Spain, along with other EU member-states, had long demanded the creation of a financial instrument exclusively devoted to the protection of the environment. Spain has also linked this instrument to the principle of 'cohesion', present in the Single European Act and other documents, which aims at bridging the economic gap between the more advanced and the less advanced countries in the EU.

Before the Maastricht Summit, the possibility of whether this new fund would be created was unclear. Finally, the treaty included the principle of cohesion as a Union protocol that would function, among other things, as a special fund for financing environmental protection measures; and, with regard to environmental issues, the treaty also introduced the principle of qualified majority for the adoption of some decisions related to environmental policy.

After the Maastricht Treaty came into force in November 1993, the Cohesion Fund lost its provisional character: for the period 1993–99, Spain will receive 7.9 million ECUs from this fund. The country will also receive 25 per cent of the budget allocated to the EU Structural Funds. All this has made Spain the principal recipient of EU structural resources. The agreement on the amount of money that the Cohesion Fund allocates to Spain should bring to a close the situation whereby the government somehow paralysed environmental policy while waiting for the arrival of EU aid, but it does not necessarily mean that the state of the environment will improve in Spain. According to a widespread rumour, only 20 per cent of the fund will be assigned to environmental measures, the rest being spent on infrastructural projects [*Ecología y Sociedad, 1992*]. Moreover, the receipt of maximum benefit from the environmental Cohesion Fund is dependent on the existence of projects integrated within a national strategy that combines socio-economic and environmental concerns [*Información de Medio Ambiente, 1992*]. Unfortunately, this strategy is currently missing.

CONCLUSION

Although the statist institutional design in Spanish pollution control policy is a factor which accounts for the implementation deficit in this area, this design enjoys a significant resilience. New political instruments like the environmental pacts have not, so far, substantially contributed to the emergence of a new atmosphere of co-operation between public agencies and industrial groups. The fact that institutional designs are the result of historical and structural factors deeply embedded in everyday political practice, helps to explain their pervasiveness. But the lack of interest, the unwillingness, or the inability of the EU to interfere in the specific implementation processes of its member-states, is also of paramount importance in this explanation.

One way in which Spain could be forced to alleviate the implementation deficit in pollution control policy would be through the Cohesion Fund. This fund examines very closely the application of its grants, since the projects approved must respond to a conscious strategy which combines economic development and environmental protection. If those projects have been deficiently designed or are badly implemented, the EU is likely to intervene in the domestic arena. This possibility allows environmentalist groups and other types of NGOs to denounce and oppose public plans on environmental grounds. The government, for fear of losing EU aid, would then be pressed to co-operate and reach agreements with private actors for securing the arrival of these resources.

NOTES

1. Pollution control policy is herein understood as air and water pollution, and waste management. Since pollution abatement is one of the objectives of environmental policy, this term will sometimes be substituted for 'pollution control policy'.
2. The focus upon industry is based on two other reasons, besides its privileged role in pollution control policy:
   - an 'objective' reason; industries are seen as one of the main groups responsible for pollution;
   - a 'subjective' reason; there is an interest in analyzing the relationship between government and industry in a suitable policy area.
3. Ireland, despite its geographical location in the north, belongs to the group of peripheral countries. Its less advanced level of economic development has converted it into an ally of the southern countries on some environmental issues.
4. No case has been found of voluntary pacts, or voluntary agreements of a global character.
5. Although member-states are the most relevant actors in the EU scenario, they are by no means equal in power, nor have they shown the same degree of interest in influencing environmental policy. Environmental proposals stem basically from three types of source:
   - endogenous and active sources, or the internal process of political deliberation basically carried out by the Commission and, occasionally, by other EU institutions (such as the European Parliament when it sends recommendations to the Commission);
   - endogenous and passive sources, or the imitative process whereby the EU copies the innovative and advanced environmental policies of some countries;
   - exogenous and active sources, or the external process whereby certain member-states deliberately try to influence EU environmental policy.

   Needless to say, when countries want to leave their mark on this policy area they use the third type of mechanism. This is generally used by countries at the centre which want, indirectly, by means of the EU, to leave their footprint on countries at the periphery [Aguilar, 1992].
6. The other member-states have played an intermediate role in the EU environmental policy.

REFERENCES

Aguilar-Fernández, S., 1992, *Políticas Medioambientales y Diseños Institucionales en España y Alemania: La Comunidad Europea como Escenario de Negociación de una Nueva Área Política*, Madrid: Institute Juan March.
Bennet, G., Weizsaecker, E.V., Baldock, D., Lavoux, T., Hannequart, J-P., Maier-Rigaud, G. and Vonkeman, G., 1989, *The Internal Market and Environmental Policy in the Federal Republic of Germany and the Netherlands* (Arnhem: Institute for European Environmental Policy).
Boletín Informativo del Medio Ambiente (BIMA), 1980, No.14, Madrid.
CIMA, 1978, *Medio Ambiente en España* (Madrid: MOPU).
Commission of the European Communities, 1990, *The VII Annual Report to the European Parliament on Commission Monitoring of the Application of Community Law 1989*, Brussels.
Costa Morata, P., 1985, *Hacia la Destrucción Ecológica de España* (Barcelona: Grijalbo).
Ecología y Sociedad, 1992, No.8, Madrid.
Información Ambiental, 1986, No.9, Madrid.
Información de Medio Ambiente, 1992, No.11, Madrid.
Kraemer, L., 1988, 'Einheitliche europaische Akte und Umweltschutz', in H.W. Rengeling (ed.), *Europaische Umweltrecht und europaische Umweltpolitik* (Berlin: Carl Heymanns).
Liberatore, A., 1989, 'EC Environmental Research and EC Environmental Policy', *Working Paper 89/407* (Florence: EUI).
Linz, J.J., 1981, 'A Century of Politics and Interests in Spain', in S. Berger (ed.), *Organizing Interests in Western Europe* (Cambridge: Cambridge University Press).
Martín Rebollo, L., 1984, 'Las Relaciones entre las Administraciones Públicas y los Administrados', in J.J. Linz et al., *España: Un Presente para el Futuro. La Sociedad* (Madrid: Instituto de Estudios Económicos).
Martínez Salcedo, F., 1989, 'La Coordinación de la Administración Central del Estado', in *Hacia una Política Integral del Medio Ambiente* (Madrid: Jornadas Trujillo, MINER).
Mayntz, R., 1978, *Vollzugsprobleme der Umweltpolitik* (Stuttgart and Mainz: Kohlhammer).
Mopu Informa, Nov. 1990 and 1991, Madrid.
Sbragia, A., 1992, 'The European Community and Implementation: Environmental Policy in Comparative Perspective', paper presented at the annual meeting of the American Political Science Association, Washington DC, August.
Streeck, W. and P. Schmitter, (eds.), 1985, *Private Interest Government: Beyond Market and State* (London: Sage).
Suárez Marcos, A., 1990, 'Visión General de la Problemática Asturiana sobre el Medio Ambiente', in Ernst & Young (aula de gestión), *La Estrategia de las Empresas e Instituciones Asturianas para el Cumplimiento de la Normativa sobre el Medio Ambiente*, Oviedo.

Environment and the State in the EU Periphery: The Case of Greece

MARIA KOUSIS

INTRODUCTION

National responses to EU environmental policy comprise a highly debated topic, since the various views entail dissimilar implications regarding the role of national states in policy implementation. Three perspectives appear to cover the various social science views surrounding EU environmental policy implementation analysis.

In the first perspective, EU environmental policy is treated as a given exogenous variable, the focus of relevant research centring upon the identification and remedy of undesired outcomes such as 'implementation deficits'. The views under this perspective draw from a 'strong European integration' source model, according to which the drive towards greater economic, political, and social integration will lead to increasingly uniform environmental policy across member-states [*Buller* et al., *1991*].

The second perspective does not necessarily treat EU environmental policy as an exogenous variable. Resting on the assumption of a 'weak European integration' source model, it challenges the nature of EU environmental policy itself, its focus embracing the dynamics of policy formulation at the centre and the possible conflicts between EU policy and national interests and goals. Arp [*1991*], for instance, addressing the issue of car emission controls, points to the catalyst role of some member-states, including Greece, in opposing the introduction and enactment of new standards. Also, it has been suggested that, while other EU policies may be contradictory as regards environmental protection, member-states, in implementing environmental policy, are seriously expected to take into account the sensitive relationship between growth and the environment (see Introduction to this volume).

The third perspective does not rest on a European integration source model, be it strong or weak. Instead, it challenges 'sustainable development', in which the EU's Fifth Action Programme on the environment is grounded, and encourages member-states to regard ecosystem protection as a guiding element and not an impediment to economic growth. This view rests on the 'ecological modernization' model which proposes technologically induced developments as a solution to ecological imbalances in the EU [*Spaargaren and Mol, 1992: 323–44*]. The model for some researchers is not mutually exclusive with that of European integration [*Buller* et al., *1991*].

Under the first and third perspectives the role of the state would be rather passive. A strong European integration model places explicit focus on EU policy implementation and enforcement, while ecological modernization restricts the role of the state simply to encouraging the industrial sector ecologically to modernize, and persuading its citizenry that technological breakthroughs are ecologically compatible. State intervention is considered as an 'unproductive' element in the long run because it tends to hinder the innovation process. The weak European integration assumption, however, allows member-states more room for discretionary action, given the discrepancies in various EU policies, the dynamics of environmental policy-making itself, and the fact that responsibility for implementation rests exclusively with member-states.

Independently of the way in which the issue is approached, variation among national responses to EU environmental policy are observed and require studying with emphasis on the role of the state, especially in the Union's periphery, within this theoretical framework. While ecological modernization is associated with countries of the core, the other two perspectives are incomplete unless they manage to unravel the role of the EU peripheral state in environmental matters.

The purpose of this chapter is, first, to present a political economy approach to EU environmental policy implementation analysis, in an effort to add to and qualify existing social science views on this subject. Second, using available literature and secondary data sources, the chapter attempts, on the one hand, to interpret the Greek experience regarding environment and the state over the past twenty years in the light of this approach, and, on the other, to place it in the above theoretical framework.

ENVIRONMENT AND THE STATE: A SOCIOLOGICAL VIEW

Regardless of the environmentally friendly face presented by capitalist producers today, capitalism greatly restricts liberal democracy and the administrative state [*Dryzek, 1992: 18–42*]. Since it requires growth to avoid political instability, it neglects the future and provides no mechanisms for dealing with problems of common property and public goods, which markets generate. This imprisonment is reflected in the limits within which environmental problems are dealt with. According to this perspective, liberal democracy is limited in promoting more progress in this area, mainly because of its skewed power distribution favouring the business world, its short time orientation, and its dependence on economic growth. The administrative state is highly constrained in its responses to these problems, given severe difficulties in policy implementation, the inability to deal with complex problems due to the non-neutrality of expertise, and its obstruction of the free transmission of information.

An extreme view of the state identifies it as a self corrupting professionalized corps of legislators, bureaucrats, and military forces, exercising power at the expense of popular power under a capitalist system that denaturalizes the physical environment [*Bookchin, 1989*].

State theories can be separated into liberal and structuralist [*Fitz-Simmons* et al., *1991: 1–16*]. The former hold a Hobbesian view of society in which all interests have equal access to power (liberal pluralism), while the latter emphasize the economic role of the state as an institution bound to enhance the conditions of accumulation [*Poulantzas, 1973*]. A distinct view, assigning the state a more dynamic and complex role, and according to which state actions are determined by autonomous state interests, is presented by Skocpol [*1980*].

Among the structuralist theories is one which views the crisis of the welfare state as a conflict between accumulation and legitimation [*Offe, 1975; O'Connor, 1988: 11–38*], and which has been applied to environmental questions as well. Humphrey and Buttel [*1982*] and Schnaiberg [*1994*] maintain that in a dynamic capitalist society the state has two main roles: a) to ensure the conditions for profitable capital accumulation and economic growth, and b) to maintain social harmony. Under the first role, the state is committed to looking at environmental resources for their exchange values, whereas under the second role it has to preserve the ecosystems' capacities so that they can produce the use values of various political constituencies of state actions.

The logic of the treadmill of production requires that ecosystem elements be converted by capitalists through market exchanges into profits. This drive for the accumulation of social surplus supplements two more characteristics of the modern treadmill: the expansion of production is shifted away from localized subsistence towards national and international markets, and the institutional apparatus generates additional demands for increased ecosystem utilization in order to strengthen and support an even greater capitalization of production [*Schnaiberg, 1994*].

Three possible syntheses exist for the societal-environmental dialectical conflicts arising during the history of expanding economic activities, according to this view: the 'economic synthesis' that stresses surplus production over environmental protection; the 'managed scarcity synthesis', which acknowledges the legitimacy of both goals for a sustained industrial state, and the 'ecological synthesis', which stresses ecosystem protection over surplus production. For the prevailing managed scarcity policies, Schnaiberg [*1994*] delineates competing claims by various parties interested in ecosystem use, in the form of conflict repertoires. He also stresses that, although the contextual factors play an important role in shaping trajectories and policy outcomes, the fiscal crisis of the state restricts its support for environmental research and more effective environmental protection.

On the basis of the above political economy view, the state's first role of facilitating economic growth entails the following. Under the treadmill of production, state bureaucrats, along with economic actors and workers, are impelled to demand even greater ecosystem utilization for further capitalization of production. In doing this, the accumulation process is promoted through transfer payments to the private sector, the necessary funds being drawn out of taxation. At the same time, however, corporations are permitted to externalize the environmentally damaging costs of production and limit corporate liability [*Cable and Benson, 1992*]. The state's managed scarcity intervention is limited by the following four different types of producer strategies: reducing initial problem consciousness, constraining implementation of environmental protection policies, restricting enforcement actions or evading controls, and increasing public and political resistance [*Schnaiberg, 1994*]. These strategies are often labelled, 'corporate environmental crimes', pointing once again to the failure of the state's environmental regulatory process due to the inherent contradictions of the liberal democratic state [*Cable and Benson, 1992*].

The failure of the state to address adequately environmentally damaging activities affecting public welfare and thereby providing for social harmony, often generates opposition in the form of grassroots environmental organizations which challenge the state's role as a legitimator [*Cable and Benson, 1992; Modavi, 1993*]. In conflicts over the implementation of state policies, state agencies and actors benefit from solutions mollifying these movements without imposing high exchange value costs on local capitalist producers [*Schnaiberg, 1994*]. Simultaneously, legitimation concerns of the polity can act as political opportunity structures for the grassroots to pressure the state and gain concessions [*Modavi, 1993*]. These arguments are strengthened by evidence that governmental agencies seek to achieve the minimum acceptable level of environmental remediation at the least economic cost, usually via non-structural technological fixes [*Gould, 1992*].

PERIPHERALITY AND THE ENVIRONMENT

Peripherality within the EU carries geographical and economic location connotations, being at the same time a designation used by EU administrative authorities [see Introduction to this volume]. The core-periphery distinction, however, is mostly used by world systems theorists such as Wallerstein [*1974*] and Gunder Frank [*1979*]. 'Semiperiphery', which includes both capitalist and socialist nations, is a useful world-systemic category for the study of contemporary issues, including the environment. For peripheral European regions the term 'semiperiphery' is more appropriate than 'periphery'. The role of semiperipheral states in environmental matters, and specifically that of Greece, can be systematically analyzed first by deriving the capitalist character of social organization in the semiperiphery and then looking at the limits of the state's role in environmental remediation, in the context of sociological theories regarding the liberal state.

Recent works [for example, *Martin, 1990*] systematically place countries, such as those of the EU periphery, on the semiperiphery of the world economy. A recent attempt to place countries in the semiperiphery of the world economy is that of Korzeniewicz and Awbrey [*1992*], who use Arrighi and Drangel's [*1986: 9–74*] classification of 102 nations' structural positions in the mid-1980s. This classification is based on wealth (long-term income), via the use of gross national product (GNP) per capita, as an indicator of the relative distribution of aggregate rewards which reflect the distribution of core and peripheral activities. Under this scheme, semiperipheral states are those where benefit appropriation exceeds the long run costs of participation without reaching the levels of wealth *standards in core countries. Semiperipheral states with stagnant economic* growth, such as Greece and Portugal, are characterized by social and political arrangements that impede their long-term upward mobility in the world economy, while geographical proximity to the core may provide for other semiperipheral states (for example, Spain, Ireland, or East Germany) the opportunities for mobility that are absent elsewhere [*Korzeniewicz and Awbrey, 1992: 609–40*].

In order to bypass the difficulties which arise from restricting definitions, Mouzelis [*1986a*] uses the term semiperiphery in an unbinding fashion to encompass societies which are characterized by late capitalist industrialization and early parliamentarianism. Greece is one of the examples he uses, given its significant development of social overhead capital during the nineteenth century which was linked with the growth of its export sector. In the post-war period, Greece managed to extend its industrialization aided by multinational capital. Thus, industry ended up contributing more to the gross national product than the agricultural sector, and there has been a marked shift in exports from raw materials and agricultural produce to industrial goods. This capitalist development is restricted, uneven, and accompanied by an overwhelming urbanization process. The industrial sector contains a few very large high productivity capitalist enterprises employing a considerable section of the wage earning force, and a plethora of small, low productivity, family-oriented units. The persistence of small, simple-commodity production units existing side by side with huge firms (usually state or foreign-controlled) exercising a quasi-monopolistic market control is a striking characteristic of the semiperiphery [*Mouzelis, 1986b*].

Greece is not only semiperipheral in its capitalist formation, but also in political terms. The country's relatively strong civil society and its parliamentary institutions are closer to those of the West than to those of Third World societies. What distinguishes Greece from core European states, however, are the clientelistic and personalistic relationships between political parties and its citizens [*Mouzelis, 1986b*]. In the post-war period clientelistic relations have been rooted in access to the state and the potential to influence the distribution of state funds and public sector jobs through kinship ties and political affiliation [*Comninos, 1990*].

Environment and the state in the EU periphery or the world system's semiperiphery is a relatively unexplored topic. It needs to be asked, therefore, whether, in this respect, the semiperipheral state resembles the core state. This means addressing the state's conflicting functions and the way in which it interacts with grassroots opposition, the ultimate question being how this hinders or facilitates EU environmental policy implementation.

ENVIRONMENT AND THE STATE IN GREECE

During the 1970s, when Greece was an associate EC member, practically no limits were placed upon the exploitation of Greek natural resources in the name of economic growth. This is clear from the relevant literature and the country's five-year development plans [*Koutsoumaris, 1976; KEPE (Centre of Planning and Economic Research), 1976, 1986*]. The environmental damage, however, especially that relating to the health of Greek citizens in the most polluted area of Metropolitan Athens, pressured the state to take action. Thus, in the first half of the 1970s, an Environmental Pollution Control Project (EPCP) was established under the auspices of the Greek Ministry of Social Services, the World Health Organization (WHO), and the United Nations Development Programme (UNDP). Further, the Greek constitution, which was passed shortly after the departure of the military government, made specific reference to the need for environmental protection and the government's direct responsibility towards that end. In both the 1970s and the 1980s, the government facilitated growth and financed various environmental protection projects in the context of planning for economic and social development.

Greece, like other EU-member countries, is incorporating all environmental directives into its national legislation. Nevertheless, the receipt of warning letters and the country's appearance before the European Court for violation of green rules indicate the existence of 'implementation deficits' [*Sbragia, 1992*]. In the following subsections, the political economy approach is used to analyze in greater detail the role of the state in environmental remediation, given that it has an organic interest in playing both the accumulation and legitimation roles.

The State as Facilitator of Growth

Since the early 1960s, as an associate member of the EC, the Greek state has facilitated growth, especially in the areas of industry, tourism, and agriculture through various economic incentives including low interest loans, subsidies, and tax allowances. This function was promoted at both the private and state producer levels, leading to greater ecosystem utilization. The Greek state not only acted as a facilitator of growth but, unlike the core states, also took over the role of major producer. This is a very important aspect which needs careful analysis. A more hybrid activity of the state, closer to its legitimation role, is its promotion and execution of environmental protection projects in the context of planning for economic and social development.

At the private producer level, the Greek state's deep desire to facilitate growth is reflected in the enactment of special legislation, usually referred to as a 'Development Act', authorizing the subsidization of private investment projects out of state money, as well as in the Parliament's occasional failure to pass an Environmental Pollution Control Act, for instance the one drafted in 1979 by the Ministries of Industry and Labour which would have regulated industrial pollution. It is believed that the failure to pass this particular act was due to the expected negative impact it would entail for an already weak industrial sector which was trying to survive the second oil crisis and the European recession of the 1970s [*Lekakis, 1990: 465–73*]. Various ecosystems, however, have been placed under threat by industrial activity, as indicated in Table 1.

This picture did not change drastically in the years following 1981, when Greece became a full member of the EC. Two Development Acts were passed during this period and the private sector has not been the target of environmental controls, especially in view of the highly publicized issue of ailing or 'problem firms'. Funded by the EC, EPCP drafted a pollution control plan for Metropolitan Athens and suggested controls limited to good housekeeping practices as regards industrial pollution – that is, saving energy and combustion products reduction – rather than asking for mandatory installation of pollution control technology [*EPCP, 1984*].

TABLE 1

PRIVATE INDUSTRY REGIONAL PROFILE AND THREATENED ECOSYSTEMS

	Total no. of plants	Highly polluting plants	Threatened ecosystem
Region:			
E Sterea & Cyclades*	5,172	617	Evoikos Bay, Larimna
C & W Macedonia	20,302	223	Salonica and Ptolemais airsheds
E. Macedonia	5,286	54	Kavala Bay
Peloponese & W Sterea	12,879	200	Patras and Korinth Bays, Pinios and Alfios Rivers
Thessaly	8,004	95	Pagasitikos Bay, Pinios River
Epirus	3,115	33	Not available
Thrace	4,005	51	Tsiflic-Benteli Streams
Crete	85	16	Not available
Isles of the Aegean	Not available	11	Geras Bay

*Excluding the Greater Athens Region
Source: KEPE, 1986.

In an attempt to cast a framework for environmental protection and curb industrial pollution without endangering growth, the government passed the Environmental Policy Act of 1986 and in the same year reached a compromise agreement with the Federation of Greek Industrialists which qualifies all anti-pollution investment projects in the Athens Metropolitan Area to a 30 per cent direct capital investment subsidy and a 40 per cent interest subsidy on borrowed capital [Lekakis, 1990]. Thus, Greek environmental policies do not rely on strict regulations for the private producers since strict environmental controls would hinder Greek industrial development. A recent survey of 170 large industries provides evidence which points to the token effort they make to minimize the withdrawals from or additions to ecosystems. More specifically, of these producers, 41 per cent stated that they did not have any relationship with environmental pollution, while 51 per cent said that they did carry out some environmental protection measures to abide by Greek, EC, or international law [Ktenas, 1992]. The European integration process will intensify competition among manufactured goods and hence Greek firms will eventually have to replace their outdated capital stock if they are to survive. This process of using modern technology may have some favourable environmental impacts, without constituting a permanent answer to the problem of industrial pollution in Greece.

At the state producer level, one of the main actors involved is the Public Power Corporation (PPC) which holds the monopoly of energy in Greece. Older electric power plants, such as those in the northern province of Kozani, or that of Keratsini in the Metropolitan Athens region, have heavily polluted their nearby residential areas and ecosystems. Planned PPC facility construction threatens ecosystems which are protected by international agreements such as the Ramsar convention, signed by the EC and Greece individually. Examples of these are the Valia Calda bear park, the hydroelectric projects on the Nestos river [*Yphantis, 1993: 27–37*] and the diversion of the Acheloos river. In all of these cases PPC's relationship with the state (the Ministry of Industry, Energy and Technology) is one of crucial importance; and, of course, in many instances, this state producer receives EU grants and financing. Recent research [*Kousis, 1993: 3–24*] has shown that the state has been very supportive of PPC activities, especially when the EC has been involved through financial assistance. In fact, this relationship is shown to be stronger than that between private producers and the state. In the name of national economic growth, PPC, backed solidly by the state (under any given ruling party), has been asking inhabitants of various rural Greek regions to assume all environmental costs in the form of environmental damage and lost regional product.

State activities centre around other issues as well. In addition to providing support for big private and state producers, central state agencies aid local government and smaller private producers. This is done either directly through public subsidies and grants or through other forms of assistance. An example of the latter is aerial spraying by the Ministry of Agriculture over 200,000 acres of olive trees in 36 provinces, and ground-based spraying of another 80,000 acres. This 'free bonus' entails many dangers, since the spraying medium contains toxic substances such as Lebaycid, which can damage the nervous systems of life forms that come into contact with it. It is estimated that 25 per cent of this medium adheres to the trees while the other 75 per cent is diffused over the areas sprayed, which include a plethora of small villages dispersed in the countryside [*Orfanidis, 1989*].

Promotion of agricultural development has also been pursued directly through various methods of support to agricultural units, especially the ones using modern greenhouse technology which receive public grants and subsidies under the national development acts in operation. Through the creation of a single internal market, protection from external competition, and financial assistance, the EC's common agricultural policy also assisted agricultural growth in Greece, although its volume and composition has been altered [*Maravegias, 1989*]. The result has been increased area-wide pollution which is a function of overproduction.

Tourism is another area heavily influenced by the state's accumulation function. In the early 1970s, Greek financial institutions provided investment financing to tourism entrepreneurs, using non-banking criteria. A recommendation by the state-controlled Hellenic Tourist Organization was sufficient and the loan was guaranteed by the state. Following the departure of the military government, financial institutions returned to banking criteria but borrowers were able to meet their loan requirements since land values increased [*Kousis, 1984*]. In the 1980s the government strongly favoured tourism growth. These policies overall promoted the construction of large hotels in rural areas where infrastructure is still inadequate. As a result, a disproportionate demand on environmental resources is exercised during the peak tourist season.

The state facilitates a growth in ecosystem utilization, while at the same time promoting environmental projects directly. During both the 1970s and the 1980s, national development plans included state environment rehabilitation projects funded through the government's Public Investment Programme (PIP). In Greece, unlike countries of the core, those projects literally placed the burden of environmental protection on the state (which is also consistent with the constitutional dictum) rather than the private sector [*Lekakis, 1991: 1627–37*]. Most of these projects concern the construction of sewerage and waste water treatment facilities. The immediate outcome of these projects was the maintenance of GNP figures at rather steady levels – they would have been on the decline otherwise – and an increase in annual low-skill low-wage employment by over one per cent. Therefore, the state rushed to select, on the one hand, measures that would not penalize the private sector, and, on the other, environmental projects that would increase national output and employment.

The growth facilitation picture of the Greek state portrays lively elements of clientelistic relationships. Access to state officials through kinship ties, political economic support and affiliation, and bribing have yielded various legal anomalies occasionally picked up by the Greek press. Suspected bribery incidents include the accusation of key Environment Ministry officials for allowing highly toxic wastes to be dumped in the sea [*Lambropoulos, 1992*]. Recent clientelistic examples concerning the misuse of biologically renewable natural resources include the extension of the harvest season for fishing techniques which threaten the fish stock in the long run, and the building of expensive villas in forested areas when this is strictly prohibited by law [*Kourmousis, 1993a, 1993b*]. Thus far, clientelistic relationships have proved a successful route for those who wish to employ non-sustainable biomass reducing technologies in the exploitation of environmental resources.

The State as Legitimator

The state maintains social harmony when it preserves the ecosystems' capacities so that they can produce the use values of various political constituencies of state actions, since environmental conflicts are defined as struggles over decisions to allocate or restrict access by groups or social classes to ecosystems [*Schnaiberg, 1994*]. Social harmony calls for state actions which constitute the operational components of its legitimation function. Three broad avenues appear to be available to the state in acting for ecosystem preservation: introduction of national environmental regulation and enforcement, participation in and implementation of international agreements, and any other state action associated with public participation or public opposition to projects or plans regarding both growth and environmental remediation.

With reference to the first avenue, the preceding analysis of the Greek state's growth facilitation function indicates, on the one hand, the state's concern over loss of business survivability in a fast changing and highly competitive world, and, on the other, the selection of environmental projects which advance rather than retard national economic growth. Further, the state's involvement as a producer inevitably generates bias in enforcing environmental controls. The remaining two avenues are perhaps more critical in explaining lags in EU environmental policy implementation.

International agreements usually reflect the mutual interest of two or more states to undertake action to protect certain ecosystems (for example, between the US and Canada over the protection of the Great Lakes), and they may be legally binding according to international law. EU directives are more or less by-products of the European integration treaty and they do not necessarily reveal the member-countries' willingness to take action. An EU directive essentially constitutes a supranational determination of the national state's legitimation role. Even in the case of specific international agreements, the signatories are not always fulfilling their commitments and hence their legitimation function [*Gould, 1992*]. Therefore, EU environmental policy 'implementation deficits' can be explained only if the two functions of the state originate from its organic interest to perform them.

The final avenue for state action in ecosystem preservation is through responses to public pressure, especially local environmental movements. Table 2 shows a composite picture of such movements during the period 1982–92. Public pressure groups operate as informal means of control and they may facilitate EU environmental policy implementation when their goal is to curb pollution from private and state producers.

TABLE 2
ENVIRONMENTAL MOVEMENTS IN GREECE, 1982–1992

Year	Number of recorded movements per area of concern*				
	Air	Water	Soil	Quality of life	Total
1982	2	1	1	2	6
1983	-	1	1	-	2
1984	-	-	-	4	4
1985	9	2	3	9	23
1986	2	4	-	3	9
1987	7	2	-	9	18
1988	1	-	-	3	4
1989	5	-	3	8	16
1990	3	3	3	1	10
1991	1	1	-	2	4
1992	4	2	5	4	15
Total	34	16	16	45	111

*Recorded in *Nea Oikologia* (New Ecology), a monthly Greek ecological magazine.

In the 1970s, lack of effective environmental controls often produced massive public opposition, and state agencies such as the Ministries of Industry and Health put pressure on the private sector to reduce industrial pollution. This pressure was exercised by invoking existing fragmentary legislation and short-term measures such as the use of cleaner fuels, etc., which never threatened business profitability [*Lekakis, 1990*].

The Ptolemais pollution problem, caused by PPC mining and electricity generating activities, soon led to the emergence of grassroots movements which acted as informal means of control and received extensive publicity in the media. Using strategies such as rallies, marches, strikes, media announcements, forming and supporting approximately one hundred ecological organizations, Greek citizens mobilized to put pressure on the state to correct the environmental injustice they experienced. Positive state actions that followed as a result of these movements included the installation of electrostatic precipitators, which give the state its legitimator role.

The closing of the Keratsini power plant constitutes another example of the state's legitimation function. During the 1970s this plant, a state producer facility located at Piraeus, was one of the major air pollution sources in the Metropolitan Athens region. Operating on low quality crude oil, the plant generated at least three times the quantity of sulphur oxides as the entire space-heating source, releasing more than a quarter of the total emission load in the area. Following intense mobilization of the Piraeus residents, in the late 1970s, protesting against what they called 'the destroyer of their health', the government ordered the closure of the plant.

During the 1980s a few strong waves of grassroots environmental protest sent their clear messages to the state. They developed strategies to combat the threat to their health, livelihood, and ecosystem, given the state's facilitation of nonlocal objectives. One of the reported incidents occurred during 1988–89, when a Northern European company filed an application with the Environment Ministry and the Ministry of Industry requesting a licence to install a toxic waste storage and treatment facility in a west-central Greek community. Initially, the company was treated positively and the ministries requested environmental impact studies before granting final approval. At that time, however, the local residents learned of the company's interest and acted immediately. They occupied a coastal installation which was the company's location target, blocked their provincial motorway, and held a public referendum which crystallized their determination never to allow the location of this or any other similar plant in the area [Kousis, 1991]. These activities 'forced' the state to reconsider the case, given its potentially damaging effects on the local ecosystem. Thus, as a result of the serious local reactions, in 1990, the two Ministries rejected the company's request. In this instance, the state viewed the positive benefits to the local residents and the national economy, which the company had promised to deliver, as less important than the negative impacts which local residents foresaw for their health and environment.

The public resistance to an EU co-funded renewable energy (geothermal) development project, proposed by the PPC on the island of Milos, is an interesting case, revealing also the conflict between EU resources and environmental policy [Kousis, 1993]. Fully supporting the PPC, the state systematically avoided local requests for careful monitoring of the geothermal plant's polluting activities, and the compilation of an environmental impact assessment study. The government and the PPC had a common goal, the continuation of the project in economic, political, and national interests, and purposely played down the potential risks from this 'environmentally benign' renewable energy source. The local population's reaction was a large organized effort involving all the island's inhabitants. They initially financed an independent impact assessment study which proposed the closure of the station and, later, using strategies such as occupation of public buildings, a long march to the station, press releases, strikes, and picketing, they finally succeeded in getting the project halted. Under this social pressure the government, as of early 1989, decided to suspend all activities at the pilot station, which is still out of commission.

The above cases are only indicative of the state's response to environmental movements which act as informal means of control. There have been cases where problem complexity has diminished its legitimation function. A typical example is photochemical pollution in Metropolitan Athens, which has repeatedly stirred fierce public opposition but cannot be addressed successfully simply by controlling producers. Furthermore, controlling producers cannot always prove effective, as the mandatory 30–40 per cent reduction in industrial output during smog emergency periods, is usually counterbalanced by an increase in the number of working shifts.

In some instances public pressure may assist in regressing rather than implementing EU policy. This occurs often in the area of liquid waste management, but predominantly in the management of solid waste, when pressure groups impede access to ecosystems by municipalities seeking sanitary landfill sites. In north-western Greece, residents from 102 villages and three towns, located within the wider Kalamas river basin, strongly resisted the disposal into that river of domestic sewage and all other wastes subjected to secondary treatment from the urban centre of Ioannina [*Kousis, 1991: 96–109*]. They claimed that such an act would seriously deplete the quality of the Kalamas water on which they depend for their livelihood. The area covers a large acreage of irrigated fields, scenic resorts for recreation, and potential sites for fish culture and the further development of tourism. From May 1987 to November 1990, and continuing through early 1993, the residents mounted an impressive show of power, not against their neighbours, but against the central government, which is responsible for subsidizing the waste water treatment project and making the final decision. Protesting residents closed down harbours, motorways, government buildings, banks and shops and attracted national attention. The result has been the halting of related construction activities until further studies are carried out considering alternative solutions.

The problem of solid waste disposal, finally, for many municipalities in

Greece, is a rather dramatic experience. The Environment Ministry, in its first annual report to the Greek Parliament, emphasizes the 'problem of social acceptability' as the most serious obstacle in establishing municipal sanitary landfill sites [*EM, 1992*]. Chanea and Heraklion municipalities in the northwestern and north-central parts of Crete respectively typify this experience. Chanea dumped its solid wastes for some years into Couroupitos (a 'free disposal' site in the Akrotiri hills) and Greece was taken to the European Court for this. Recently, local authorities in the area have blocked the route to Couroupitos for the city's solid waste transportation vehicles. At the same time, while efforts to build a sanitary landfill facility have met the stern opposition of the Akrotiri people, the city has symbolically protested against the government by surrounding the Prefectoral building with its solid waste vehicle fleet. Heraklion used a solid waste disposal site within the territorial boarders of Fodele community, under a contract. Fodele is currently refusing to renew the contract and a conflict between these two local governments has developed, with the Department Governor (Prefector) threatening to take action, including the use of police, if Fodele does not agree to a six month renewal until a permanent solution is reached.

Thus, looking at legitimation, on the basis of the evidence provided herein, the state's hybrid activity of undertaking directly the financial burden to execute projects aiming to restore ecological balances, is supplemented by occasionally positive responses to public demands for a cleaner environment. However, opposition to environmental remediation projects often engineered by local authorities remains a perplexing problem and a policy issue which certainly warrants additional research.

CONCLUSIONS

Semiperipherality in the world system is a more appropriate characterization of peripheral EU countries. The evidence drawn from this study highlights certain features which distinguish Greece from core nations of the world. Specifically, the Greek experience shows how labour-intensive environmental projects of the 1980s were engineered to support higher GNP and employment levels. It was also inferred that the tougher competition anticipated within and outside the EU will eventually lead to a replacement of obsolete manufacturing capital stock and hence to reduced industrial pollution. This opportunity, however, does not necessarily indicate that ecological modernization can serve Greece or the EU as a model for environmental policy, especially since it does not favour state intervention in environmental remediation.

In Greece, the state possesses the two conflicting roles of growth facilitator and legitimator, which the political economy perspective ascribes to the liberal democratic states of the core. The fragile industrial structure, however, in which national planners have historically invested a lot, helps in explaining the state's reluctance to introduce strict and costly environmental regulations. Growth versus the environment materialized as a sensitive issue when high technology environmental protection projects were to be undertaken by the manufacturing sector which would threaten its competitiveness. Thus, while in the more industrialized nations, producers are more or less inclined to adopt environmental controls and pass the costs on to the consumer, in the less industrialized ones the state assumes the larger share of environmental investment costs. The semiperipheral state, however, appears closer to the core state in its will to legitimize its citizenry. The strong civil society in post-war Greece has been pressuring the state to fulfil its legitimation function.

Environmental policy implementation in a semiperipheral state such as Greece, however, is conceived as being hindered more than in the core state, firstly because the presence of state producers, solidly backed by the state, enhances its growth facilitation function, and secondly due to social phenomena such as clientelistic relationships, which various segments of Greek society resort to. Finally, policy implementation is affected positively or negatively as a result of public opposition to projects contemplated by private and state producers. These features definitely tend to support a view whose source is a weak, rather than a strong, European integration model.

ACKNOWLEDGEMENTS

I wish to thank the editors of this volume and an anonymous referee for their useful comments on an earlier draft.

REFERENCES

Arp, H.A., 1991, 'European Community Environmental Policy: What to Learn from the Car Emission Regulation?' Paper presented at the seminar, 'European Integration and Environmental Policy', Woudschoten, The Netherlands, 29–30 Nov.

Arrighi, G. and J. Drangel, 1986, 'The Stratification of the World Economy: An Exploration of the Semiperipheral Zone', *Review*, Vol.10, pp.9–74.

Bookchin, M., 1989, *Remaking Society* (New York: Black Rose Books).

Buller, H., A. Flynn and P. Lowe, 1991, 'National Responses to the Europeanization of Environmental Policy: A Selective Review of Comparative Research', Paper presented at the seminar, 'European Integration and Environmental Policy', Woudschoten, The Netherlands, 29–30 Nov.

Cable, S. and M. Benson, 1992, 'Acting Locally: Environmental Justice and the Emergence of Grassroots Environmental Organizations', revised (Sept. 1992) version of a paper presented at the 1992 meeting of the American Sociological Association, Pittsburgh.

Comninos, M., 1990, 'Local Dimensions of the Clientelistic System', in M. Comninos and E. Papataxiarchis (eds.), *Community, Society, and Ideology* (Athens: Papazisis [in Greek]).

Dryzek, J.S., 1992, 'Ecology and Discursive Democracy: Beyond Capitalism and the Liberal State', *Capitalism, Nature, Socialism*, Vol.3, No.2, pp.18–42.

EM (Environment Ministry), 1992, *Annual Report to the Greek Parliament on the Environment* (Athens, November [in Greek]).

EPCP (Environmental Pollution Control Project), 1984, *Summarized Action Programme to Combat Pollution and Traffic Congestion in Athens*, Memorandum to EEC, Feb.

FitzSimmons, M., J. Glaser, R. Monte Mor, S. Pincetl and C. Rajan, 1991, 'Environmentalism and the Liberal State', *Capitalism, Nature, Socialism*, Vol.2, No.1, pp.1–16.

Gould, K., 1992, 'Legitimating Growth: The Role of the State in Environmental Remediation', paper presented at the 1992 meetings of the American Sociological Association, Pittsburgh.

Gunder Frank, A., 1979, *Dependent Accumulation and Underdevelopment* (New York: Monthly Review Press).

Humphrey, C.R. and F.H. Buttel, 1982, *Environment, Energy, and Society* (Malabar, Fla: Krieger Publishing Co).

KEPE (Centre of Planning and Economic Research), 1976, *The Environment*, A report for the five-year Development Plan of Greece, 1976–1980, Athens (in Greek).

KEPE (Centre of Planning and Economic Research), 1986, *Protection and Rehabilitation of the Environment*, a report for the five-year Development Plan of Greece, 1983–1987, Athens (in Greek).

Kourmousis, M., 1993a, 'Tarachi gia ta gri-gri' (Turmoil over gri-gri fishing boats), *Eleftherotypia* (Free Press), 20 Jan. (in Greek).

Kourmousis, M., 1993b, 'An Ehis . . . Ladosi tha to Sosis' (If you have . . . bribed you will save it), *Eleftherotypia* (Free Press), 16 March (in Greek).

Korzeniewicz, R.P. and K. Awbrey, 1992, 'Democratic Transitions and the Semiperiphery of the World Economy', *Sociological Forum*, Vol.7, No.4, pp.609–40.

Kousis, M., 1984, Tourism as an Agent of Social Change in a Rural Cretan Community, PhD Dissertation, Department of Sociology, The University of Michigan.

Kousis, M., 1991, 'Development, Environment, and Mobilization: A Micro Level Analysis', *The Greek Review of Social Research*, No.79, pp.96–109 (in Greek).

Kousis, M., 1993, 'Collective Resistance and Sustainable Development in Rural Greece: The Case of Geothermal Energy on the Island of Milos', *Sociologia Ruralis*, Vol.33, No.1, pp.3–24.

Koutsoumaris, G., 1976, *Financing and Growth of Greek Manufacturing* (Athens: Institute of Industrial and Economic Studies [in Greek]).

Ktenas, S., 1992, 'Anti na epivarinthi to kostos . . . na epivarinthi to perivallon' (Place the Burden on the Environment and not Production Costs), *Vema* (Tribune), 8 Nov. (in Greek).

Lambropoulos, Th., 1992, 'Apovlita YPEHODE . . .' (Wastes from the Environment Ministry . . .), *Eleftherotypia*(Free Press), 21 Aug. (in Greek).

Lekakis, J.N., 1990, 'Distributional Effects of Environmental Policies in Greece,' *Environmental Management*, Vol.14, No.4, pp.465–73.

Lekakis, J.N., 1991, 'Employment Effects of Environmental Policies in Greece', *Environment and Planning A*, Vol.23, pp.1627–37.

Maravegias, N., 1989, *Greece's Accession to the EC and its Effects on Agriculture* (Athens: Foundation for Mediterranean Studies [in Greek]).

Martin, W.G. (ed.), 1990, *Semiperipheral States in the World Economy* (New York: Greenwood Press).

Modavi, N., 1993, 'The Political Economy of State Intervention in Land Use Conflicts: A Case Study of Community Opposition to Golf Course Development in Hawaii', *Case Analysis*, Spring/Summer.

Mouzelis, N., 1986a, *Politics in the Semiperiphery: Early Parliamentarism and Late Industrialization in the Balkans and Latin America* (New York: MacMillan).
Mouzelis, N., 1986b, 'On the Rise of Postwar Military Dictatorships: Argentina, Chile, Greece', *Comparative Studies in Society and History*, Vol.28, No.1, pp.55–80.
O'Connor, J., 1988, 'Capitalism, Nature, Socialism: A Theoretical Introduction', *Capitalism, Nature, Socialism*, Vol.1, pp.11–38.
Offe, C., 1975, 'The Theory of the Capitalist State and the Problem of Policy Formation', in L.N. Lindberg et al. (eds.), *Stress and Contradiction in Modern Capitalism* (Lexington, Mass.: D.C. Heath).
Orfanidis, C., 1989, 'Aerial Spraying against Dakos', *Eleftherotypia* (Free Press), 4 June (in Greek).
Poulantzas, N., 1973, 'The Problem of the Capitalist State', in R. Blackburn, ed., *Ideology in Social Science* (New York: New Left Books).
Sbragia, A., 1992, 'European Community and Implementation: Environmental Policy in Comparative Perspective', paper presented at the annual meeting of the American Political Science Association, Washington DC, August.
Schnaiberg, Allan, 1994, 'The Political Economy of Environmental Problems: Consciousness, Conflict, and Control Capacity', in Dunlap R. and Michelson W., (eds.), *Handbook of Environmental Sociology* (Westport, CT: Greenwood Publishing).
Skocpol, Th., 1980, 'Political Responses to Capitalist Crisis: Neo-Marxist Theories of the State and the Case of the New Deal', *Politics and Society*, Vol.10, No.2, pp.155–201.
Spaargaren, G and A.P. Mol, 1992, 'Sociology, Environment, and Modernity: Ecological Modernization as a Theory of Social Change', *Society and Natural Resources*, Vol.5, pp.323–44.
Wallerstein, I., 1974, *The Modern World System* (New York: Academic Press).
Yphantis, S., 1993, 'Nestos: A River in the Sights of the Public Power Corporation,' *Ecotopia*, 21, pp.27–37 (in Greek).

The European Union and the Visegrád Countries: The Case of Energy and Environmental Policies in Hungary

JANNE HAALAND MATLÁRY

INTRODUCTION: THE VISEGRÁD COUNTRIES, THE EU AND THE CASE OF HUNGARY

This chapter deals with a part of the EU's periphery that is only lately coming to be considered as a sub-region – the Visegrád countries. These countries, named after the town in Hungary where their heads of state regularly meet to co-ordinate foreign and other policy, make up a unit through which they have concluded so-called 'Europe-agreements' with the EU, that stipulate trade and other bilateral accords as a preliminary step towards future EU membership [*de la Serre, 1991; Kramer, 1993*]. The group, in effect a sub-region of east-central Europe, consists of Poland, the Czech Republic, Slovakia and Hungary. Implicit in the Europe-agreements is that the Visegrád countries must adapt to EU policy, including environmental policy. This is a major driving force in the forging of environmental policy in these countries, as they are all very eager to join the EU, and all lack a developed environmental policy since this issue-area was almost totally neglected in the Communist period.

Thus it is interesting to ask to what extent and in which ways the EU influences emerging environmental policies in this sub-region. In order to examine this question it is necessary to concentrate on selected aspects of it because, both within the EU as well as in the sub-region itself, the development of an environmental policy which is integrated with other policies is a process which is currently only in its infancy. In the EU, environmental policy has been a 'competence' only since the signing of the SEA (Single European Act) in 1986; in the sub-region there was little or no environmental policy during the forty years of Communist rule. The period to be studied thus dates only from about 1989–90 for the sub-region, and from about 1987 for the EU.

In addition, the issue area of environmental policy resists exhaustive definition. Its subject matter derives from the normal economic and other activity in a society, and is thus essentially connected to various other policy areas. The salient question thus becomes to what extent environmental demands are made criterial for policy-making in these 'normal' policy areas, especially in the cases where there is a conflict between environmental policy and the issue area in question. In the EU, such conflicts are typified by, for example, the energy/carbon dioxide tax that has been proposed, an issue where strong conflicts of interests exist. In this case, economic policy is concerned with relative competitiveness between the EU and other trade areas, while environmental policy is concerned with how to effect changes and savings in energy use. Conflicts commonly arise when environmental policy criteria are imposed on other policies.

In east-central Europe the importance of environmental policy on the national policy agenda can fruitfully be studied by analyzing the state's ability to impose such criteria in the face of opposition. I thus propose to examine, as an indicator of the viability of environmental policy, the degree to which it is integrated with other policy. I expect the degree of integration to be more advanced in the EU than in east-central Europe since an approach aimed at integration is much more difficult and demanding than a regulatory approach, typified by – for instance – fines on emission levels. An 'integrative' approach is a new one even in western industrialized countries.

It is to be expected that the EU's aim of creating an integrated energy-environmental policy will to some extent be reflected in its policy towards the sub-region. However, since such a policy is at this stage relatively rudimentary even in the EU itself, there is not a well developed policy towards east-central Europe in this regard. Also, the historical newness of environmental policy as such in the region makes it less probable that there is much actual integration of environmental criteria in other policy areas.

With these caveats in mind, the approach of this chapter is to concentrate on one type of environmental policy and to look at how well environmental criteria are integrated with normal political activity in a given issue area. I propose to look at energy policy, and specifically at the extent to which environmental criteria are being introduced into Hungary's emerging energy policy, though the chapter's conclusion will deal with the sub-region's overall adaptation to EU environmental policy.

The general orientation of Hungarian foreign economic policy is conditioned by the Europe-agreement with the EU. It stipulates the requirements in the economic field for eventual membership, and entails a gradual adaptation of Hungarian domestic policy to EU regimes. As we shall see, in terms of energy-environmental policy proposals, adaptation to EU rules in, for example, coal policy is a stated goal. Despite the fact that the Europe-agreement has been extremely tough on the question of freer trade, particularly in agricultural products [*Tovias, 1991; van Ham,*

1993], it remains the only route to EU membership, which is the stated goal of all Hungarian policy-makers. Compliance with the terms of the agreement in major policy areas like the environment is therefore critical.

Hungary has been in the forefront of adaptation to economic policies of the West, and has already privatized large parts of its industries. Also, Hungary has advanced perhaps the farthest in trade with the West, and in introducing market mechanisms for pricing, including energy pricing. This makes it the more interesting as a case to use for the purpose of evaluating EU influence on energy-environmental policy, for in the EU itself there is a similar process going on. Wherever there is a contradiction between environmental policy concerns and market concerns in the energy sector, there needs to be a political ability to intervene. Such a political capacity is evolving, albeit very slowly, in the Commission [*Weale and Williams, 1993; Matláry, 1993*].

I plan to analyze the emergence of energy-environmental policy in Hungary in the period from 1989 and also the simultaneous development of the EU's policy towards the region of east-central Europe in this regard. In order to establish that the latter has had an influence on the former, signs of similarity alone are insufficient. Evidence from interviews and other sources would be needed to show that this was in fact the major driving force. I can identify direct influences in, for example, the form of institution-building aided by EU funds, and in the employment of advisors from the Commission in the various Hungarian ministries, but conclusions about the direct influence of the EU must nonetheless, in my view, remain tentative.

In the following I shall first analyze EU policy instruments towards east-central Europe as well as the integration of EU environmental policy with energy policy. There follows an analysis of Hungarian energy policy and a discussion of Hungarian environmental policy and institutions as they pertain to the energy issue area. Finally an assessment of the EU's role, present and potential, is made.

EU ENERGY-ENVIRONMENTAL POLICY TOWARDS EAST-CENTRAL EUROPE

The acute need for more and cleaner energy in eastern Europe has acted, along with the trend towards a stronger degree of common energy policy in the EU, to accelerate the merging of energy and environmental policy. Three policy areas are important in evaluating the role the EU has played and may play in the effort to press for cleaner energy use in Europe; first, the degree to which the EU has a sufficiently strong energy policy to impose rules that take care of environmental concerns; second, the importance that environmental concerns actually play in EU policymaking, especially as regards energy policy; and third, which policy instruments exist in relation to east-central Europe. Here I shall only focus on these elements to the extent that they bear directly on the EU's role towards east-central Europe in the issue area.

The SEA provided a formal basis for environmental policy in the EU, while the TPU (Treaty on Political Union) in several ways strengthened the obligation on the part of the EU to integrate environmental policy with other policy areas [*Wilkinson, 1992; Baldock* et al., *1992*]. However, despite this obligation, so far there has been little attempt at such integration. The one example of integration of energy and environmental policy, the energy/carbon dioxide tax, has met with considerable resistance [*Matlary, 1993*]. One must therefore conclude that this is a policy process that is so far only in its infancy, but in which the legal instruments and the mandate in the TPU are significant, and in which the Commission seeks to influence the integration of environmental criteria also in so-called 'Third Countries' such as Hungary.

The EU has several formal political ties to the region of eastern and central Europe [*Pinder, 1991*]. For aid for economic development, there is the European Bank for Reconstruction and Development (EBRD). The underwriters include the EU, US, and non-EU countries in Europe. Situated in London, the Bank started its work in April 1991 [*Europe*,[1] 1 June 1990]. The EBRD operates as the western financial instrument for eastern Europe, and has been instructed to give priority to projects that improve the environment [*Europe*, 30 Jan. 1992] 'The environmental reflex' must be incorporated from the start, according to the European Commission [*Europe*, 11 Jan. 1991]. The contribution from the EBRD in 1992 for specific projects in the region was 621 MECU (million ecu) [*Brewin, 1993*]. Investment in the region amounted to 2.7 billion US dollars in the same year, all sources included.

Further, the EIB (European Investment Bank) gives loans also to the energy sector and is interested in favouring projects that are environmentally sound. From 1991 it was authorized to lend in Poland and Hungary as well as in the former East Germany, and later to other countries in the region [*ECE*,[2] June 1990]. Thus, a loan of 50 MECU has been granted to Poland in order to modernize its gas industry. The money is earmarked for desulphurizing Polish natural gas, thus contributing to a cleaner environment [*ECE*, Aug. 1990]. In 1991, 93 per cent of its loans were granted to energy-related projects.[3] In 1992, the EIB granted a total of 1.7 MECU to the region [*Kramer, 1993: 22*].

In terms of political participation, both the Commonwealth of Independent States (CIS) and the eastern and central European countries will be part of the European Environmental Agency (EEA), an EU institution agreed in 1990 and to be situated in Copenhagen. It is expected to begin operating in 1994, and will monitor compliance with and implementation of environmental policy in the participating countries [*ECE*, June 1990]. Further, the Europe-agreement with the Visegrád countries stipulates that they participate fully in all EU environmental programmes.

The EU has also contributed to the funding of the Regional Environmental Centre in Budapest, whose main task it is to assist NGOs (non-governmental organizations) in the region in their development, act as a meeting-place for information and educate local communities in environmental management and implementation.

Specific EU programmes focus on improving air and water quality in these countries [*Europe*, 30 May 1990]. In its selection of projects in 1990, the PHARE programme (Poland and Hungary Action for Restructuring the Economy), set up to assist in the development of eastern Europe [*PHARE, 1992*], gave 'overwhelming priority to those related to environmental protection' [*Europe*, 17 May 1990]. In 1990 alone, 47 MECU were given to environmental measures in Poland and Hungary. PHARE is by far the most important programme; Hungary received 15 MECU from it in 1991 and 10 MECU in 1992 for energy and environmental projects. Most of this goes towards establishing an administrative infrastructure for utilizing environmental charges and taxes, and to the funding of pilot projects aimed at energy saving and efficiency [*ECE*, June 1992].

PHARE has been extended from the original end date of 1992 until 1997, and now relates to ten countries in eastern Europe. Yugoslavia was excluded in 1991. It has an OECD-constituency, but it is the Commission that co-ordinates the work of the so-called G-24 group, which is made up the donor countries with the IMF, the World Bank and the EIB as observers. PHARE covers all aid to 'reconstruction' and emphasizes privatization. Its 1992 budget was 1014 MECU. The energy and environmental areas figure prominently in this context. PHARE is a sector-specific programme, which also establishes offices in the recipient countries. Thirteen energy offices, including one in Budapest, have been opened. The intention of the latter is to provide energy management experience and technological know-how on site.

The G-24 deals with energy and environmental problems in eastern Europe in general and the emergency character of the energy supply situation in particular. It adopted a declaration on economic assistance to the region in February 1990 [*Europe*, 22 Feb. 1990]. The EU Commission, which co-ordinates the work of this group, proposed that energy financing become a priority in light of the double stress under which these countries had come, resulting from the demand on the part of the CIS for energy payments in hard currency, and the Gulf crisis that threatened to cause increases in the oil price. The Commission stressed the need for a 'medium-term energy strategy on a *pan-European scale*' (my emphasis), proposing that all financial instruments be co-ordinated, that help be given in diversifying dependence on Russian energy on the part of eastern and central Europe, and that assistance in developing alternative gas import sources be provided. The reception of the Commission proposal by the Group was mixed, and no immediate measures were taken [*Europe*, 31 Oct. 1990]. However, the importance of this meeting lies, not in its results, but in the nature of the Commission's proposals. The substance of these proposals indicates that the Commission takes on the responsibility for co-ordinating and developing a full-fledged strategy in the energy-environmental area not only for its members, but specifically also for the eastern and central European region. In terms of financial policy-instruments, both the EIB and the EBRD are in place. In its support to eastern Europe in 1990, the EIB looked specifically for environmental soundness of projects.

There are, in addition, programmes that entail direct EU funding: THERMIE and SAVE. Both these programmes are part of the Commission's environmental policy, especially identified as part of its policy towards fulfilling the climate obligation of stabilizing carbon dioxide emissions at 1990 levels by the year 2000 [*Gerini, 1992*]. THERMIE concerns energy efficiency, SAVE deals with energy saving, and both programmes are part of the Commission's Fourth Framework Programme for Research and Development, adopted in 1993, which extends to certain external countries, including Hungary.

Summing up, the EU clearly commands a variety of financial and institutional policy instruments in energy-environmental policy towards east-central Europe. However, these are managed by various parts of the Commission and not tied to any unified EU policy based on environmental criteria.

THE ENERGY-ENVIRONMENTAL NEXUS IN HUNGARY: THE LESSONS OF THE PAST AND THE DEMANDS OF THE PRESENT

What we mean by energy-environmental policy in this chapter refers to the environmental problems that arise from energy use. These are mainly connected with emissions from such use. Major environmental problems in Hungary that stem from the energy sector can be summarized as: wasteful energy use in general accounts for high carbon dioxide emissions; an old car fleet accounts for considerable emissions of oxides of carbon and nitrogen in the major cities; and coal fired power plants contribute to sulphur dioxide emissions. Hungary's dependence on coal is, however, the lowest of the countries in the sub-region, only 21 per cent compared with Poland's share of 76 per cent. This represents a major advantage in terms of environmental effects of energy use. Nevertheless Hungary, as a former Soviet satellite, shares in a legacy of the past that is largely responsible for the present problems in the energy-environmental nexus.

Some environmental problems are common to the former 'satellites', and these derive from the ideological support in Marxism for 'conquering nature' by large-scale and heavy industrialization, which in Hungary led to absurd industrial developments in energy-intensive industries such as steel, and to rather bizarre agricultural ventures including rice-growing. The effect of this policy on the environment was drastic; heavy industry is energy-intensive and a major source of pollutants [*Salay, 1990; Knabe, 1989*]. Energy policy was developed in blueprints like the megalomaniac Gabcikovo-Nagymaros dam project, to which I shall briefly return, and in the policy of nuclear power plants that were built throughout the entire region. In addition, Soviet oil and gas was exported in barter arrangements with the 'satellites', and flowed very freely indeed throughout the 1960s and 1970s. The 'price' paid for oil in this period was much lower than the world price; however, this is not an accurate comparison as all energy was bartered. It was a highly asymmetric barter arrangement since the Soviets could dictate the terms. They had a geopolitical reason for covering their satellites' demand for energy: as interdependence was physical, the power of control was unquestioned. The slightest sign of political instability would induce the Soviets to supply the energy needed [*Kramer, 1992*]. The same supply lines could also act as a very real threat; cut-offs in oil and especially gas have a long history as a Soviet political weapon. Ironically the oil pipeline between the former USSR and the satellites was named the 'Friendship' pipeline, and the gas line was named the 'Brotherhood' line. The result of the long-term energy trade relationship between Hungary and the Soviet Union was a total and one-sided dependence on Soviet energy as well as an influx of abundant energy that encouraged extreme waste. The fact that this energy was never priced made cost considerations, as well as saving, irrelevant.

The result of this is that today the level of energy consumption is much higher than it need be; it is generally considered to be between 50 and 100 per cent higher than in the West. This demand must now be covered with imports paid for in hard currency at world market prices. In the residential sector, many houses lacked (and still lack) meters; cooling was achieved by opening the window.

The Soviet rationale for encouraging such a wasteful energy policy was twofold; first, their own inability to put a price on energy, and hence the consequence that the barter system was the only way of trading with the rest of the bloc – desirable goods from Hungary were demanded in return for the energy. This was unproblematic as long as the dependence was so total. Second, energy policy within the satellites themselves served as an instrument of very concrete (and primitive) political control.

The legacy of the past in the entire region can thus be characterized as one that *encouraged waste of energy*, took *no notice of pollution as a result of energy use*, and which *imposed a political ideology of heavy industrialization* which had little natural basis in countries such as Hungary. In addition, in the countries where coal could be mined, this was encouraged as a major feature of socialism; the coal-miner was an exemplar of the true proletariat.

Energy policy problems are thus legion. Imports from the former Soviet Union of both oil and gas are less and less dependable while, as of 1990, they have also had to be paid for in hard currency at world market prices. The most severe problem is, however, that the oil and gas supply is very unreliable from this source: production and transmission problems have become increasingly evident, and the political problems caused by transit through the Ukraine (which is demanding higher payments for transmission services) makes reliance on this source of energy a precarious one.

For all these reasons Hungary would like to diversify its import structure, but the alternatives are not promising. There is no oil available from the Adria pipeline, running from the port of Rieka, as long as there is war in the Balkans. Hungary ordered two mtoe (million ton of oil equivalent) of oil through this pipeline in late 1991, which is delayed indefinitely. Further, in that year Russia supplied only 4.5 mtoe of an order for 6.4 mtoe [*EEER*,[4] March 1992]. Oil is available on the world market, but very expensively.

Further, the infrastructure in the case of Hungary is such that the dependence on Russian energy will prevail for many years. Alternative transport for gas is through an extension of the pipeline from Algeria through Italy and further through ex-Yugoslavia, but this project is naturally also postponed due to the war. A Hungarian move has, however, been to extend its gas line to Baumgarten in Austria, a 114 kilometre link that will allow for imports from the West [*EEER*, Aug. 1992].

These structural parameters need to be taken into account when discussing the possibilities of integrating environmental criteria into energy policy. They clearly impose constraints on the energy choices of Hungary: there is a need for increased imports of energy, even allowing for the fact that the country's only nuclear plant in Paks will double its capacity in the near future. However, imports from Russia are insecure, both in terms of deliverability and quantity, and other sources are not forthcoming very quickly. Hungary thus does not have much freedom to choose the energy source. Even if natural gas may be preferable for environmental reasons, its supply is restricted. The only indigenous source of energy that can be argued to have few immediate environmental consequences is nuclear-generated electricity, but this is a very problematic line of argument in itself.

A policy document from the Hungarian Ministry of Industry and Trade [*1992*] outlines energy-environmental policies. Not unexpectedly, the emphasis is on energy saving and efficiency, an area where much can potentially be gained and where there are few political costs in terms of disagreement. Given the trend towards decreased indigenous production and a host of difficulties in procuring reliable imports, this is a sound step. However, the means of achieving efficiency are basically market incentives, which generally mean higher consumer prices. The pace at which Hungary is introducing market prices is very rapid; by 1992 most energy prices had been liberalized. Also, the energy sector is being privatized quickly: the Oil and Gas Trust was privatized in late 1991; coal mines are in the process of privatization, the national electricity board (MVMT) was in 1991 made into a joint stock company with the state as the major share-holder; and the regional gas companies are currently preparing for privatization [*EEER*, Aug. 1992].

However, the introduction of market pricing for energy to both industry and consumers may hold severe political costs. For instance, in October 1990, the petrol price was suddenly increased due to the Gulf War and short supplies of oil, and this almost brought down the government and left the fragile democracy on the verge of collapse. The city of Budapest and all major roads in the country were blocked for several days by angry taxi and truck drivers. Interviews I conducted at the time indicate they took the opportunity to attempt to bring down the government and to discredit democracy as such. This very grave episode that could easily have succeeded in a coup went largely unnoticed in the West, perhaps because it was seen merely as a protest against higher oil prices. It should, however, serve to remind of the brittleness of the new democratic institutions in Hungary. In terms of energy implications, this event brought home the crucial importance of finding the right balance between introducing market prices for energy and retaining political stability. Market pricing in this case meant a steep increase in petrol prices overnight in a political culture completely unused to this. This type of increase may be justified on an economic logic but is almost impossible to absorb in terms of the political cost.

This need for balance is further illustrated in the case of the reduction of coal subsidies. The Hungarian government, eager to comply with EU rules of coal subsidies, has launched a very radical programme for closing uneconomical pits: 'The development of the power sector provides the basis for the integration in the EC' [*Hungarian Ministry of Trade and Industry, 1992: 11*]. Environmental reasoning is ostensibly also behind this, but cannot be assumed to account for much as an isolated factor. However, the above policy document utilizes this very well, putting it bluntly that Hungarian power plants are 'obsolete', being 40 years old; 'their environmental pollution exceeds the European standards several times' [*Hungarian Ministry of Trade and Industry, 1992: 10*].

Over only two years, from 1990 to 1992, the number of jobs in coal mining was cut from 50,000 to 30,000, and several pits were scheduled for closure in 1993. In the same period the number of pits fell from 35 to 20. The major policy line now is that only domestic coal that can compete with imported coal will prevail. This is indeed a radical approach that compares with British coal policy in terms of rigour. Better quality coal is imported, and existing power plants are planned to be retrofitted with combined cycle turbines which allow for the use of gas in addition to coal in electricity generation. The Kelenföld plant in Budapest is an example of this process, having attracted funding from the World Bank and the European Investment Bank [*EEER*, Jan. 1993: 13]. The running down of the coal sector, however, must be presumed to have large-scale social-political repercussions, as whole communities are being eradicated. There are strong reactions from the trade unions, which claim that the 1993 closures will result in an additional 15,000 jobs lost [*EEER*, Dec. 1992]. In December 1992, a general coal strike was narrowly avoided. The policy ahead is to part-privatize hard coal mines and gradually phase out brown-coal mines; brown coal is the most polluting and carries the lowest calorific value. In the EU, coal policy is currently being revised and funds exist for developing alternative employment in former coal-mining regions. With EU policy still pursuing coherence, Hungary's radical policy should prepare it carefully to be fully consistent with future EU coal policy.

Hungary is thus caught on the horns of a dilemma: the legacy of the past in terms of an outmoded, heavy-industrial structure demand excessive energy use, and future and much-needed growth will mean increased energy use as well. The environmental NGOs often argue that energy saving will do the job and that an increase in energy use is avoidable. However, this argument relies on a no-growth philosophy that has no substantial support in the populace. After forty years of communism, materialism as a basic worldview is predominant in the young generation, and this is naturally coupled with a strong desire to achieve a degree of material wealth so long rendered impossible. No-growth arguments thus have little appeal in themselves as a political programme, although many NGOs advocate them. This is, as will be discussed later, a potential problem in making the NGO community responsive to 'realistic' environmental political strategies. Salay puts the problem rather politely thus: 'The gap between the often very theoretical and complex approach of green politics and everyday concerns of the Hungarian populace is sometimes too great to bridge' [*Salay, 1990*]. The problem is, however, two-fold: many NGOs may be rather utopian in approach, yet the populace is largely uninformed about environmental problems that affect them. A major policy strategy is therefore one of education, which is also stressed in the energy policy document in a rather charming way: 'The protection of the environment . . . requires that the public be given clear and correct information' and 'the selection of the acceptable solution, based on majority opinion, [is] necessary'. However, although the greatest possible consensus is sought, we are warned that 'it would be unrealistic to assume that there is an energy policy which can be acceptable by everyone' [*Hungarian Ministry of Trade and Industry, 1992: 8*]. In all my reading of energy policy documents, I have never come across one that is concerned with due democratic process in the sector. However, it is a valid point that education of the public is needed for practising energy saving and 'raising environmental consciousness'.

Energy use can thus safely be assumed to rise in the coming years, although the closing of much obsolete industry is causing all economic indicators to turn down at present. There seems, however, to be agreement on the fact that 1992–93 presents the bottom in this respect [*Inotai, 1992*]. The policy document on energy policy estimates that economic growth will be three or four per cent per year in a low and high scenario respectively in the period 1990–2000, but that energy demand will largely be curtailed by an increase in energy efficiency that is between 0.3 and 3.5 per cent per year [*Hungarian Ministry of Trade and Industry, 1992: charts 7 and 8*]. The ability to achieve such an ambitious energy efficiency effect will, however, depend on extensive success in introducing both the market mechanism in energy policy and in the conversion to modern state-of-the art industrial technology.

Thus, the options for energy policy in terms of both security of supply and environmental consequences offer no easy solutions. Environmentally friendly natural gas must largely come from the unstable Russian supplies through the equally unstable Ukraine for many years to come. Oil is the fuel of the transport sector and is almost all imported at world prices. Nuclear energy may be considered environmentally the best solution as it carries no emissions, but the Soviet-type technology in the region is not without its problems and a new and additional nuclear plant will be very expensive. Coal is the 'worst' energy source in terms of emissions, and is being run down very rapidly, yet its demise requires additional imports of other energy costing hard currency.

Environmental problems in this context thus can be expected to be rather peripheral to the policy-makers' agenda. Their solution can be hoped for, but realistically they must be treated as part of the larger policy nexus where they appear. By this I mean that the energy-derived environmental problems will probably not be attempted to be solved in isolation from the larger economic and structural problems themselves.

ENVIRONMENTAL POLICY-MAKING AND INSTITUTION-BUILDING IN HUNGARY

The communist era was marked by a basic and thorough disregard for environmental problems. While this policy area gradually made its way into the political agenda in the West from the beginning of the 1970s, it was largely neglected in eastern and central Europe. Apart from being, for ideological reasons, a 'non-existent' problem in communist countries, modern equipment like scrubbers or cleansing technology for industrial production cost a great deal. In the Soviet energy sector itself the conventional estimate is that both production and transmission technology lags 20–30 years behind its western counterpart, and this situation was allowed to prevail in the primary sector of the Soviet economy, justified by the pursuit of foreign currency! It is small wonder that environmental problems were ignored.

In addition to this came the factor of secrecy and a general lack of public awareness of pollution as a problem. There was naturally no public discussion of such negative aspects of industrial society, and against this background it becomes eminently understandable that almost all the ecological protest groups in eastern and central Europe developed from general oppositional groups. In other words, they evolved on a platform of environmental policy but constituted a general protest against 'the system' [*Matláry, 1991*].

This was the case in Hungary with the so-called Danube Circle and the by now infamous case of the Gabcikovo-Nagymaros dam. In the battle against the development of the Danube into a gigantic hydroelectric scheme were born both the Hungarian NGOs and the Ministry of the Environment. This case therefore merits some examination. The dam project has a long history, dating from the 1950s, when the Soviet Politburo dominated the discussions about a joint Czechoslovak-Hungarian construction of a large hydroelectric plant on the Danube which divides the two countries and also forms the international border between them. It must be kept in mind that the territory on the Slovak side, including Bratislava, was Hungarian until the Treaty of Trianon. The national question is therefore close to the root of the dam-project, while the environmental consequences are very real also. These were to include the actual diversion of the river itself for 31 kilometres, loss of the fresh-water supply for millions of people, possible contamination of the underground fresh water supply, which is the largest in Europe, disturbance of flora and fauna with repercussions for agriculture in the area, and, finally, physical destruction of one of the most valuable historical areas in Hungarian history [*EEER*, Nov. 1992]. Against this the Slovaks argued that investment there has already been too much to cancel the project, and that the energy needs served were to be substantial.

There is a long and bitter saga on this project, from its adoption by the Hungarian government in 1977 to its cancellation in 1989. The major NGO in the mobilization of Hungarian popular, and eventually governmental, protest against the dam, was the Danube Circle. This made it a prime symbol of the new democratic politics and the importance of environmental policy. The role of environmental policy therefore figured prominently in political party platforms before the first free elections in 1990. However, as in most west European countries, this policy field is the most difficult to define in practical terms and the one that lends itself the most readily to speeches on festive occasions. The 'green' parties in the elections received little support although one parliamentary committee is now concerned with environmental policy. This general picture may be no better or worse than that which obtains in the average west European polity.

Hungarian environmental policy thus exists, but its implementation is regarded as highly ineffective. Fines for polluting only punish after the misdeed, and the preventive aspects are far harder to instil than 'oldfashioned' methods of fining the trespassers. Salay concludes that there is no efficient system of control and follow-up at the regional level, and points out that monitoring is based on an 'honour' system whereby enterprises are supposed to report on themselves [*Salay, 1990: 25*]. This overall conclusion is supported by the newly formed Environmental Management and Law Association (EMLA) which states that 'there is a conspicuous lack of the institutional support, oversight, and legal structures required to implement environmental policy . . . there is a perception that economic development must take precedence over environmental protection' [*EMLA, 1992: 1*]. According to Szigeti Bonifert, managing director of the association and former environment ministry director, there is an acute need for professional management practices in environmental policy implementation lest environmental policy disappear completely from the political agenda (personal interview). The mission statement for the organization further reads that 'there is limited regulation, minimal enforcement, and inadequate experience in the spheres of environmental management and law' [*EMLA, 1992: 2*], to an extent that an altogether novel approach is needed to deal effectively with environmental issues. This association largely comprises the top Hungarian professionals in law and administration in this field. Its founding is symptomatic of how serious the lack of sufficient institutionalization of environmental policy is felt to be among the elite.

However, a new environmental law was proposed in draft form in late 1992, and more than 270 institutions and NGOs were asked to comment on the draft. This consultative process is being institutionalized by the creation of a public forum where hearings on the law will be held. In a thorough and comprehensive comparison of EU and Hungarian environmental legislation presented to the Commission by the EMLA, Bandi et al. find that there are relatively few discrepancies. The major problem does not lie in the absence of legal rules in the various environmental areas, but in the lack of policy priorities and in implementation:

> The environmental enforcement system in Hungary is far from being satisfactory due to several reasons from the anomalies of economic development in the past to several reasons embodied in the legal or organizational system. One of the most surprising handicaps of the present environmental enforcement system is that in the third year of transition into a rule of law system and a market economy there is no sign of an environmental policy or strategy. The necessary harmonization of economic and environmental interests is not a primary target of the present policy-makers [*Bándi et al., 1993: 202*].

There are some indicators of integrative thinking in the energy-environmental area, as have been identified in the policy document on energy. However, the role of environmental concerns is generally considered to carry very little independent weight in the political process. The above study in fact regards 'the need for international co-operation and the need to join the EC . . . (an) opportunity to apply pressure to address environmental degradation' [*Bándi et al., 1993: 56*]. The authors point out that although Hungary has signed a number of international environmental agreements in the past, in only one case has it enacted implementing legislation [*Bándi et al., 1993: 35*]. This is an indication of the lack of integration of environmental concerns in general in the national political process, and contributes to the impression that both in terms of planning as well as in terms of implementation, the role of environmental policy does not yet amount to very much. With regard to the energy sector, we have seen that actions undertaken in order to introduce commercial criteria, like coal and gas privatization and market pricing for energy, most likely will have benign environmental effects in terms of energy saving and fuel switching. However, these effects are the results of a general economic policy that in essence has nothing to do with the integration of environmental criteria into other policy areas. Conflicts of interests have thus not (yet) occurred.

Summing up, this part of the chapter has developed an analysis of environmental policy and its management in Hungary. The latter had a 'flying' start as the dam project united the forces concerned with, among other things, environmental issues. However, the NGO community experienced both a loss of political importance after about 1990, and problems with its ideological orientation compared with the pragmatic policy-making of the establishment. The official policy-makers, exemplified by the Ministry of Environment, face a tall order indeed, and critics contend that the administrative-legal apparatus is largely inadequate to cope with the type of problems that stem from the energy sectors and others. The EMLA even contends that environmental policy may be disappearing from the political agenda.

However, as was evident in the previous discussion of energy-derived environmental problems, solutions are being sought largely by the use of the market mechanism in the very basic sense that market pricing will discourage wasteful energy use and bring down brown coal use. In addition to this, some form of energy and/or carbon tax will eventually present itself as a policy option. So far the Ministry of Trade and Industry, responsible for energy policy, relies on increased energy efficiency as the key to reducing air pollution.

THE ROLE OF THE EU IN HUNGARIAN ENERGY-ENVIRONMENTAL POLICY

If it thus remains a relatively sound conclusion that there is yet little evidence of the integration of environmental policy into other policy areas in Hungary, this is neither surprising nor necessarily cause for rebuke. It is, after all, a new policy area, and a new way of thinking about environmental policy. Further, it is an issue area that may be assumed to demand considerable learning for politicians, for the public at large, and for bureaucrats. This has been the experience in western countries, and the learning process there is far from finished.

The integration of EU environmental policy with other policy areas is, as mentioned earlier, only in its infancy. EU policy-makers are eager to develop an environmental policy that relies on market instruments, and this trend is also recognized by Hungarian policy-makers. They propose such instruments as the key to energy efficiency and saving. However, there is probably the familiar problem of the convert being more Catholic than the Pope here; the energy policy document that sets out the new energy-environmental policy for Hungary reveals an unjustified fear of political management of this policy: 'the energy policy developed on a market basis cannot undertake to provide a detailed programme for the participants in the energy market, who operate in corporate form, since that would go back to the centralized control of the past' [*Hungarian Ministry of Trade and Industry, 1992: 11*]. This unfamiliarity with political management in a capitalist economy is not surprising, but is potentially a serious political problem in environmental policy and allied areas. As the EMLA underlined, there is in their view a real danger that the latter will be treated as a residual category that will largely take care of itself as a consequence of the general introduction of the market mechanism in energy and other policy areas. In other words, a lot hinges on the ability to integrate political institutions and policies in the environmental area with 'mainstream' politics. This is, however, the present challenge to the EU itself, not only to Hungary.

The potential role of the EU in forging an energy-environmental policy in Hungary is thus much greater than its present role. The consolidation of the EU's own environmental policy will take place in the time ahead as the TPU has come into force, while the various policy instruments towards east-central Europe are in place and have been employed for a year or two already. On the Hungarian side we may assume that the desire to join the EU will increase the political willingness to adapt to its environmental policy as well as to create conditions for effective implementation. Implementation will probably remain the major problem, as it is in many EU countries in this area, and the political process of integrating environmental criteria in other policy areas can also be assumed to present many future political conflicts. However, in both these regards Hungary is a position no different from that of other European countries.

NOTES

1. The periodical *Europe: Agence internationale d'information pour la presse*.
2. The periodical *Energy in Europe*.
3. Information given by the bank's managers at the Lisbon energy conference, 'Energie et Cohesion Economique et Sociale', 4–5 June 1992.
4. The periodical *East European Energy Report*.

REFERENCES

Baldock, D., G. Beaufoy, N. Haigh, J. Hewitt, D. Wilkinson and M. Wenning, 1992, *The Integration of Environmental Protection Requirements into the Definition and Implementation of Other EC Policies* (London: Report from the Institute for European Environmental Policy, July).

Bándi, G., S. Fülöp, M. Nagy, J. Szlávik and K. Lambers, 1993, *Environmental Law and Management System in Hungary: Overview, Perspective, and Problems* (Budapest: Report by the EMLA to the EC Commission, June).

Brewin, C., 1992, 'External Policies', *Journal of Common Market Studies*, Annual Review of Activities, Vol.31, No.3.

EMLA, 1992, *Mission Statement*, Budapest: EMLA.

Gerini, M., 1992, 'Actions dans la cadre de la politique énérgetique communautaire', paper presented at EC energy conference in Lisbon, June.

van Ham, P., 1993, *The EC, Eastern Europe, and European Unity: Discord, Collaboration, and Integration since 1947* (London: Pinter).

Hungarian Ministry of Trade and Industry, 1992, *The Hungarian Energy Policy* (Budapest: Ministry of Trade and Industry).

Inotai, A., 1992, 'Current State and Future Prospects for Economic Transformation in Hungary', paper given at the conference 'The Current State and Future Prospects for Political and Economic Transformation in East-Central European Countries', Vienna, 3–4 Dec.

Knabe, H., 1989, 'Glasnost für die Umwelt: Zur Lage des Umweltschutzes in Ungarn', *Osteuropa*, Vol.7, pp.633–48.

Kramer, H., 1993, 'The EC's Response to the New Eastern Europe', *Journal of Common Market Studies*, Vol.31, No.2, pp.213–44.

Kramer, J., 1992, 'Energy and Environment in Eastern Europe', paper presented at the European Consortium for Political Research conference in Heidelberg, September.

Matlary, J., 1991, 'Energibruk og miljøproblemer i Øst-Europa: mulige implikasjoner for EFs energipolitikk', *Internasjonal Politikk*, Vol.48, No.2, pp.293–9.

Matlary, J., 1993, *Towards Understanding Integration: An Analysis of the Role of the State in EC Energy Policy, 1985–1992*, Oslo: Centre for International Climate and Energy Research (CICERO).

PHARE, 1992, *Assistance à la réconstruction économique des pays d'Europe centrale et orientale*, Luxembourg: Information Office of the EC.

Pinder, J., 1991, *The European Community and Eastern Europe* (London: Pinter).

Salay, J., 1990, 'Environmental Management: Current Problems and Prospects', *Report on Eastern Europe*, September.

de la Serre, F., 1991, 'The EC and Central and Eastern Europe', in L. Hurwitz and C. Lequesne (eds.), *The State of the EC: Policies, Institutions, and Debates in the Transition Years* (Harlow: Longman).

Tovias, A., 1991, 'EC-Eastern Europe: A Case Study of Hungary', *Journal of Common Market Studies*, Vol.29, No.3.

Weale, A. and A. Williams, 1993, 'Between Economy and Ecology? The Single Market and the Integration of Environmental Policy', in D. Judge (ed.), *A Green Dimension for the EC: Political Issues and Processes* (London: Frank Cass).

Wilkinson, D., 1992, *Maastricht and the Environment* (London: Report from the Institute for European Environmental Policy).

Ups and Downs of Czech Environmental Awareness and Policy: Identifying Trends and Influences

PETR JEHLICKA and JAN KARA

INTRODUCTION

Roughly since the beginning of the 1970s, and particularly as a consequence of the 1972 Stockholm Conference, environmental issues have been gradually acquiring prominence, reaching a peak – for the time being – at the Earth Summit in Rio de Janeiro in 1992.

The same period also witnessed a growing variety and divergence in approaches towards the environment. While for a number of states, at least in the 'developed' part of the world, the beginning of the 1970s marked a turning point, other countries remained more or less 'frozen', allowing (or even further promoting) the continuity of the dreadful processes of over-exploitation of natural resources, careless consumption of the 'global commons' and extending their 'borrowing from the future' [see *Kara, 1992*]. Typically, this held true for central and eastern European countries, and the Czech Republic (the western part of former Czechoslovakia and from January 1993 an independent state) was no exception.

After 1989 with all its revolutionary changes, it seemed that the main barriers to the introduction of environmentally sound practices had been removed and it was believed generally that high expectations would be rewarded soon. Environmental concerns became firmly built into post-1989 politics, thanks to the active role of environmental activists in the earlier anti-communist movement and to the apparent 'idealism' of some of the new leaders. New environmental institutions have emerged, as well as far-reaching strategies and programmes.

Nevertheless, the situation was extremely difficult and the implementation of all these ideas had to face severe constraints as the whole society found itself not only in the new international setting, but also in the process of substantial reform of all aspects of life. There were new political and economic systems bringing massive privatization, taxation reform, social differentiation and unemployment, designed to usher in a new 'western-style' country as soon as possible. The environment was not deemed worthy of sacrifices and this was perfectly recognized by the winners of the 1992 election. However, the new geopolitical settings (including the influence of the EU) with their important environmental dimension seemed to serve as a stabilizing factor in this respect; they have not allowed the 'pendulum' to swing back fully.

What follows is a brief account of the period between 1989 and the present, an attempt to assess and 'periodize' this episode. Our case study of the Czech Republic illustrates relations between environmental awareness and action, as well as between domestic and international environmental politics in the contemporary European context.

INTERNATIONAL SETTING

Looking at a map of Europe, the geopolitical location of the Czech Republic is ambiguous at first sight. Despite being a non-member of the EU, no other non-member country (except for Switzerland) is geographically closer to the economic and political core of the Union. This fact, combined with exceptional historical experience and traditions, distinguishes the Czech Republic from other European post-communist countries. From an economic point of view, the old industrial region of the Czech lands represented the most developed part of the Austro-Hungarian empire and was among the most advanced in inter-war Europe. Keeping in mind the inter-war experience of a democratic society, one could say that, at the end of the communist regime, society in the Czech Republic still showed certain features of a developed society mainly in cultural and educational terms, while at the same time being economically clearly a developing nation.

The post-1945 Soviet dominance over central Europe had far-reaching consequences. The extensive development of heavy industry within the climate of the Cold War, when the growth of production at any expense was a political and military strategy, resulted at the end in unprecedented large-scale devastation of the environment. The 'heart of Europe' was dirty. The iron curtain was quite effective in breaking traditional cultural and economic bridges and converting what used to be the crossroads of European civilization into remote peripheries, but it did not prevent the increasing pollution exchange; polluted winds could not stop blowing nor the contaminated rivers flowing [*Kara, 1992: 192*].

The application of Marxist ideology in practice led to environmental devastation in all communist countries, but its low point was probably reached in the Czech Republic and in several of its regions in particular:

> Levels of urban pollution are high, especially as a result of inferior fuels used by obsolete and inefficient combustion systems and heavy industry. Most rivers and lakes are seriously polluted and are no

longer suitable as sources of drinking water. Cities and towns have inadequate water treatment facilities, many none at all. Municipal and industrial solid waste disposal is unsatisfactory and its management and techniques are outdated. There is a substantial lack of information as regards the location, volume and toxicity of dumped waste . . . The cumulative effects of these developments on man and habitat have been alarming and appear to be even more serious than was first thought . . . They concern mortality rates, life expectancy, bronchitis, lung diseases, allergic diseases, general sickness rate etc. [*Commission of European Communities, 1991: 1*].

Present environmental policy has to deal with the heritage left by the old regime and therefore it is of decisive importance to understand what the reasons were for the complete failure of the communist environmental policy. Moldan [*1990: 9–11*] distinguished three major post-war stages in the development of environmental policy:

a) Up to 1960: rapid economic development without any tackling of environmental problems;
b) 1961–1970: slowing down of economic development, the first signals of serious, although still local, degradation of the ecological balance; the first, usually misguided, attempts to apply ecological measures;
c) 1971–1990: continuing economic growth with several periods of stagnation; continuing effort to maintain a plausible level of standard of living at the expense of large-scale devastation of the environment.

After a brief survey of communist environmental policies we shall focus mainly on the post-communist era which is, from the point of view of environmental policy, divided in two different phases:

d) November 1989 to the middle of 1992: the initial very high level of public environmental awareness declining in the course of time, with hectic preparation of environmental law, setting up of environmental administration and emphasis on international environmental cooperation;
e) The middle of 1992 up to the present: low public environmental awareness and a new 'pragmatic' approach to environmental policy.

THE INEFFECTIVE ENVIRONMENTAL MEASURES OF THE COMMUNIST GOVERNMENT

The first environmental law – the Nature Protection Act – was accepted in 1955. Other legal norms followed in the course of the 1960s: the Forest

Act; the Act on the Health of the People; the Air Pollution Act. Taking this moderate legal environmental activity into account, one can hardly argue that the communists completely ignored environmental issues. There were some relatively progressive laws, for example the Water Act, but the number of exceptions granted from the Act rendered it virtually ineffective. The fulfilment of an economic plan at any expense was the only yardstick of success. Any environmental law, no matter how progressive and strict, operated under extremely unfavourable conditions. Firstly, nature and the environment were for a long time perceived by the ruling party either as things having no value whatsoever or, if they were taken into account, referred to as disturbing factors which could only impede the march towards the communist paradise. Secondly, environmental information was classed as top secret. Thirdly, there was a system of collective responsibility in operation, which meant, in fact, the total absence of integration and responsibility. Given the complex character of the environmental issues, this fact proved to be truly devastating.

During the communist era, no ministry of the environment was established. Responsibility for the environment was dispersed among other ministries. Consequences even worse than those which could be caused by the non-existence of a central body in charge of environment, were brought about by the 'guarantee system' which was implemented in the environmental sphere in 1970 [*Zeman, 1992: 9*]. Ministries were responsible for the state of those aspects of the environment with which they were somehow economically or administratively associated.

To conclude, communist environmental policy was utterly ineffective, not due to lack of law, but because of lack of enforcement. The explanation can be found in the ideology which subordinated any environmental considerations to the fulfilment of economic targets, in a concentration on the material aspects of life, and, quite contrary to declarations, in the exclusion of the people from participation in decision-making.

ENVIRONMENT AS A MEANS OF POLITICAL MOBILIZATION UNDER THE TOTALITARIAN REGIME

To explain the role of environment and environmentalists in the 1989 November upheaval, both the socio-political context and the cultural traditions of Czech society have to be considered. Some factors, such as the popular pre-war custom of spending weekends in the countryside[1] and the overall popularity of tourism, were conducive to the origin of positive attitudes towards the protection of the countryside and the environment. The first groups of nature conservationists were established in the freer atmosphere of the 1960s. Since then the environmental movement has passed through many changes, but the essential feature of its anthropocentric orientation has never been questioned.

As everybody could see the impact of the deepening environmental degradation, the communist regime could no longer hold the stance that the existence of environmental groups and their claims and objections were illegitimate. If some environmental organizations (such as the Czech Union of Nature Protectionists [CUNP], Brontosaurus, the Slovak Union of Protectionists of Nature and the Countryside) functioned on a legal basis, although under strict control, other groups were completely illegal.

Although the official organizations were relatively well supported (for example, CUNP had about 25,000 members in 1989), their work was not very effective. In general, all environmental movements and unions were intentionally kept in an atomized state and international isolation by the communist authorities. On the other hand, their 'active cores' (local groups with specific activities) co-operated with activists from illegal groups. Both sides benefited from this collaboration. Figures from unofficial groups could achieve some of their aims by means of legal organizations, whose activity became more effective, as activists from the other side were often highly qualified experts.

Ivan Dejmal, the former Czech minister of the environment, gave a clear-cut explanation of a strong political appeal of the environment for the general public in an interview for the weekly *Respekt* [*Lamper* et al., *1993: 9*]:

> We have to realize that before November 1989 interest in the environment was a substitute. Many other issues were hidden behind it. In the last few years of the communist regime the environmental protests were, in fact, a semi-legal form of expression of differing opinions. According to the constitution, the Communist Party was a leading force in the state holding total power and, by virtue of this fact, also total responsibility. When people pointed out this responsibility, the party could not point to them as agents of western intelligence services. Additionally [from 1970 onward], citizens escaped to their privacy . . . This climate was suddenly broken by the attack of the devastated environment. One had a car, but one had to drive it on congested roads and breathe terrible exhaust fumes, because catalysers existed only in newspapers . . . It did not matter that people could buy more food than Poles and Romanians if they knew that food was contaminated. Thus, the regime poisoned the only pleasure that it offered for people. Consequently, ordinary people saw the environment as a major issue. That was the second reason

why it was a substitutional problem. People were concerned with the degradation of the environment merely due to their personal discontent, either political or consumerist. There was no mark of deeper, philosophical aspects that we must not damage the Earth.

Since environmental activities were not completely illegal, more people could take part in them than in the case of 'pure dissidents' dealing with human rights. This is not to say that human rights defenders and environmentalists were entirely separate. For example, in 1983, the Ecological Section of Biological Society prepared for Charter 77 the 'Report on the Ecological Situation in Czechoslovakia', a sharp diagnosis describing the extent of environmental catastrophe in Czechoslovakia. Thus, at the end of this period, everything was prepared: analyses and diagnoses, personalities and programmes. It was broadly known what was wrong with the environment and – more or less – what should be done and where. This was a big advantage in the time of political take-over. However, this was also a source of bias; the concepts prepared under the cover of the totalitarian regime were not always realistic enough to become immediately applicable under the circumstances of the transition. This in some cases led, directly or indirectly, to difficulties and tensions within the governing structures. Nevertheless, the period immediately following the November revolution was unprecedentedly favourable for environmental concerns.

ENVIRONMENTAL POLITICS AND PUBLIC AWARENESS AFTER NOVEMBER 1989: TWO PHASES

'Enthusiastic' Period (November 1989–mid-1991)

The appeal of environmental issues appeared from the very beginning of the political transition period. Nothing documented it better than the fact that the first party which was being established spontaneously in many places was the Green Party [*Jehlicka and Kostelecky, 1994*]. By the end of November, local nuclei of the Green Party existed in all big cities. Yet by the end of the vote in the following June, the number of supporters had dropped to four per cent. As all opinion polls conducted during the first post-upheaval phase to March 1991[2] continually showed respondents giving a high priority to environmental issues, the Greens' electoral failure could not be interpreted as a lack of environmental concern among the population. Other reasons, among them the low credibility of the Czech Greens, who were often called 'melons' (outside green, inside red) loomed behind this result.

The high standing of the environment among political issues was reflected in the manifestos and pre-electoral campaigns of all the major parties which, later on, entered the parliaments. The environmentalists who entered the political arena during the upheaval represented, later on, the electoral winners – Civic Forum in the Czech Republic and Public against Violence in Slovakia. One of the most characteristic features of the 'enthusiastic' phase was the close relationship between state officials and activists from non-governmental organizations (NGOs). The unusual influence of NGOs on the government's environmental policies was embodied in the origin of the Green Parliament in autumn 1990. The Czech Ministry of the Environment initiated the setting up of the 'parliament', composed of representatives of almost all NGOs, as its consultative body.

The results of the 1990 November local elections reflected lasting and even increasing environmental concern among the population in the most devastated north Bohemian area. For example, in the heavily polluted city of Decin with a population of over 50,000, the Green Party won 24.4 per cent of the local vote [*Jehlicka and Kostelecky, 1992: 78*].

The Period of Fading Environmental Concern (mid-1991 onwards)

The following year, 1991, witnessed not only the end of the Civic Forum, but also the next turning point for the environmental cause in the Czech lands. The politics of social consensus emphasizing such values as civil rights, democracy, sustainable development and environmental protection, was changed for a more standard and pragmatic climate of political rivalry and the defence of group interests. Politically, the overall change was embodied in a move from loose, 'under-one-roof' movements towards a more standard and clearer party structure. The practical application of what had been merely verbal slogans about market economy, liberalization of prices and privatization, from the beginning of 1991, led to profound changes in the system of values and attitudes of large groups of the population. New opportunities on the one hand were accompanied by growing uncertainty and anxiety on the other. This development gradually resulted in people's concentration on personal and concrete everyday life problems and shifted aside the 'more abstract' issues which one could not solve, like the environment.

Economic reform was set in motion in January 1991. The initial shock from the increase of prices by 26 per cent on average was followed by monthly increases of about two per cent. Further macro-economic development was relatively successful. The annual rate of inflation in 1992 was 10.7 per cent in Czechoslovakia [*Sujan, 1993*]. In Sujan's view, the depth of economic recession (the lowest national product, lowest industrial output, the highest share of unemployment) was reached in the last four months of 1991. Since then stabilization and moderate recovery has started. In December 1992 the level of unemployment in the Czech Republic was 2.5 per cent.

The results of the 1992 June election confirmed the country's support for rapid economic reform and the almost all-embracing attachment to material aspects of life. The electoral outcomes resulted in changes of personnel in the Ministry of the Environment and other state environmental institutions and a new 'pragmatic' period of environmental policy was thus institutionalized. A different approach by the new central authorities towards environmental problems could be anticipated from the manifesto of the election winner – the Civic Democratic Party (CDP). The essence of the party's approach to the environment could be perceived from one sentence: the basic principle of the CDP's environmental programme was the idea that environmental quality depended on economic prosperity. At the same time, the manifesto stressed the long-term goal of gradual compliance with EC environmental standards.

However, the economic and materialist explanations do not wholly account for the loss of interest in the environment. Firstly, going back to Dejmal's substitution argument, the environment lost its political attraction. Secondly, the advocates of the environment did not prove to be very effective. The Green Party has never come out with any measure or policy and NGOs refused to co-operate with it. Czech NGOs remained very much anchored in the old patterns of activity; concentrating on nature protection and refusing to accept non-active, but fee-paying members. They did not adapt to a new and more effective model, with large memberships and professional managers and experts.

Various opinion polls provided clear evidence that environmental issues were no longer of primary concern, but had fallen in relative importance to a position ranging from fourth to eighth.[3] After the 1992 general election, the relationship between state environmental administration and NGOs underwent a profound change. Step by step, many 'first wave' activists working in the state environmental administration were dismissed and the new authorities no longer considered the voice of the NGOs worth listening to.

At the same time, in parallel with the weakening of environmental groups, their primary opponents – economic and industrial interest groups – started to regain ground. This particularly applies to the 'energy lobby' which has been very successful in influencing government's decisions. They succeeded in persuading the central authorities that (regardless of the extremely high energy consumption per unit of production) the further expansion of the energy sector and the growth of its production are inevitable for future economic development.

ENVIRONMENTAL POLICY AFTER NOVEMBER 1989: TWO PERIODS

The 'High Tide' of Environmental Policy (November 1989–June 1992): New Environmental State Administration and Law

The first practical step in the field of new environmental policy – the setting up of the Ministry of the Environment of the Czech Republic – was taken in the wake of political take-over in January 1990. This Ministry, established with wide authority, signalled the end of the highly ineffective and sectorally divided old environmental protection arrangements. The first action of the new Ministry was the publication of the 'Blue Book', an analysis of the state of the environment in the Czech Republic, followed by another book – the 'Rainbow Book' – a programme of the Ministry's future activity. The next step was the 'State Programme of Protection of the Environment for Czechoslovakia' (approved in April 1991), pointing out major areas of concern and problems to be covered by projects funded on a grant basis by the government. The Programme also set guidelines for relevant state institutions, a strategy for legislative work and environmental policy development and proposals for handling environmental research, education, regional environmental problems and international co-operation.

After the June general election, the Federal Committee for the Environment was established. As the republic authorities administered environmental policy in their territories, the new federal authority focused mainly on legislative, conceptual and international aspects of environmental policy. The idea behind this was not to create another ministry, but a collective body composed of ministers of the environment, deputy ministers of foreign affairs, finances and economy and also chairmen of committees of the environment of all three parliaments.

After its establishment in June 1991, the Czech environmental inspectorate gradually took over inspections of water management and air pollution. Its most important task is to halt further and disproportionate pollution of the environment and to bring about strict compliance with all laws, regulations and decisions taken by state authorities.

The other important environmental institution – the State Environmental Fund of the Czech Republic – started to operate in October 1991. Payments for waste-water effluent, air pollution charges, charges for dumping waste and fines imposed by the inspectorate represent part of the Fund's income. Its expenditure supports investment and non-investment activities, scientific research and new technologies favourable to the environment.

In addition to institutional reform, there was much new legislation. The goal of post-November environmental legal activity was not filling holes in the old system of law, but creating a new type of environmental legislation which would react to new needs and approximate the EU's legal norms.

The first such law passed by the federal parliament was the hitherto-lacking Waste Management Act, in May 1991. The Act represents a set of complex legal measures based on the principle of producer-responsibility and imposing a duty to recycle waste and implement waste-free technologies. This law, together with the second federal environmental act – the Air Pollution Act – provided a basis for laws consequently passed by the republics' parliaments. The Czech National Council (as well as the Slovak National Council) passed a whole range of more specific environmental laws derived from the new federal norms, including a State Waste Management Agency Act, a Nature Protection Act, and an Environmental Impact Assessment Act.

There were also economic instruments for environmental policy, based on a variety of theoretical premises. First of all, economic tools were designed to have both punitive and stimulative effects. Secondly, although these principles were generally applicable, the scale of sanctions and stimuli were to be adjusted to the concrete conditions of a particular region. Thirdly, the system of economic tools had to be open and capable of further development.

In short, the first 'high tide' wave brought considerable achievements in developing preconditions for improvement of the environment, even if a really comprehensive and future-oriented environmental policy (strategy), based on broad integration of environmental policy into other sectors, remained a vision.

An integral part of the general strategy towards sustainable development was a positive attitude to international co-operation. Several environmental conventions were ratified or signed in 1991 (the Basel, Espoo and Ramsar conventions); at the same time significant progress was made in developing new international (both bilateral and multilateral) linkages. New inter-governmental agreements were endorsed with a number of west European countries as well as the EU, strengthened by the Association Agreement at the end of 1991.

The 'Low Tide' (June 1992 onwards): A Period of Integration?

After the June 1992 election environmental activists returned to their pre-1989 radical criticism. From their point of view, the first round of the 'environmental battle' had been lost. They may have had a point, but on the other hand one could have stressed the opportunities afforded by the new legislation and increases in the enforcement of environmental provisions. In fact, many of the achievements of the previous period had been firmly incorporated into the system of decision-making and administration and continued to operate. Total environmental investment, composed of all expenses on projects aiming at reduction of pollution levels (for instance the costs of building sewage plants and of introducing environmental technologies) maintained its growing tendency till 1992.

Some problems can, however, be identified. The State Environmental Fund was supposed to derive income from payments and fines for polluting, but these were generally set low and often not collected at all because of the insolvency of culprits. The fund therefore remains tightly dependent on the state budget.

Environmental concerns are recognized in the system of taxation introduced in January 1993. Even when a majority of original proposals was rejected (by the Ministry of Finances), some of them were incorporated into the new VAT (value added tax), consumer and income taxes. A reduced five per cent VAT tariff (the normal tariff is 23 per cent) is applied to recycled paper, biogas, car catalysers or electric vehicles; a consumer tax is levied on hydrocarbon fuels and in the newly conceptualized income tax the possibility of six years tax holiday applies to some environmentally-friendly activities, for instance the operation of alternative power generators or biogas production.

The splitting up of Czechoslovakia has had particularly serious negative implications in the area of international environmental co-operation. Here, unfortunately, the continuity of some activities has been threatened, and in some cases (including international assistance agreements) the 'momentum' has been entirely lost due to the hasty dissolution of federal bodies, the limited capacity of their republic counterparts, and the general preoccupation with the issue of Czech-Slovak separation. Indeed, the interministerial negotiation about environmental measures served, among others, as a battlefield in the fight for Slovak national self-determination.

Although the ongoing economic reform has been successful so far from a macroeconomic point of view (with low inflation and unemployment), it has not brought positive effects in terms of economic structure. Zeman [*1992: 18*] pointed out the increase in the production of energy which occurred already in June 1991 compared to the same month of 1990, although the total industrial output dropped by 32 per cent over the same period. This fall was only slightly manifested in metallurgy while the output of other industries dropped substantially, which meant an undesirable trend towards the relative strengthening of heavy industry. Czech companies have found out that they can easily sell products with a very low technological standard or bare raw materials. As their prices are relatively low they are able to compete if exported a short distance. The result is that the Czech Republic exports huge amounts of cement, timber, steel and rolling-mill materials.

The orientation of foreign investment made a substantial contribution to this trend. An effort to attract foreign capital often made the Czech authorities approve its admission where the production should rather have been reduced due to its excessive energy and material consumption. So far, foreign capital has preferred to enter plants producing building materials, mainly cement and lime works and also heavy manufacturing and electro-industry [*Stoklasa, 1992: 21*]. At the same time, the attention of foreign companies concentrates on liability for the old environmental damage. There are two different opinions about this problem. Some suggest new owners should fulfil the old obligations, an approach which would discourage many foreign investors. The second view holds that the state, as a former owner, is responsible for the old environmental sins. This approach passes the risk of high costs on to the state, but at the same time the growth of foreign investment can be expected and economic recovery should enable the state to finance environmental remedies.

The enforcement of environmental policy is considerably impeded by an inadequately developed state environmental administration at district and municipal level. The administrative structure itself, with elected bodies only on the municipal level, and without any regional self-government, makes dealing with regional environmental problems very difficult. This is compounded by the lack of environmentally qualified officers on a municipal level.

ENVIRONMENTAL CO-OPERATION OF THE EU AND THE CZECH REPUBLIC

The Association Agreement

The Association Agreement between Czechoslovakia and the EU (then EC), signed in December 1991, dealt in one of its articles (Article 80) with the environment. Despite its broad formulations, the Article touched on almost all areas of co-operation. It was assumed the Article would serve as a basis for further more specific agreements and programmes [*Cuth, 1992: 233–4*]. As to areas of co-operation, monitoring and information systems were mentioned in the first place followed by the fight against regional and trans-boundary air pollution. Furthermore, the Article covered various areas from energy saving, protection of drinking water sources, waste, and the protection of forests and species from planning and development.

An exchange of information, experts and the development of information systems were seen as being of primary importance among forms of co-operation. Harmonization of environmental law, regional and international co-operation and the development of global strategies were additional items on the list of forms of co-operation.

Poland and Hungary Action for Restructuring the Economy: the PHARE Programme

Financially, a great deal of co-operation designed in the Association Agreement was expected to be covered by the PHARE Programme. This programme, originally aimed at assistance to Poland and Hungary, soon expanded to cover other east European countries. From the beginning, the environment was among the high priority targets of the programme. The Czech Republic received aid from PHARE before even signing the Association Agreement. However, the amount of money aimed at environmental projects in the Czech Republic should not be over-estimated. In 1990, the overwhelming majority of PHARE aid for Czechoslovakia (30 out of a total of 35 million ECU) went for environmental projects. The next year, this share declined sharply: 5 million ECU out of total PHARE aid of 100 million. In 1992, virtually no PHARE money supported environmental measures in Czechoslovakia and the same went for the Czech Republic in 1993. Thus, the total amount for the environment between 1990–1992 was 35 million. Moreover, this amount was divided into two parts, Czech and Slovak, and therefore the actual PHARE means supporting environmental projects in the Czech Republic was even lower. Summing up, the total PHARE environmental assistance to the Czech Republic was approximately equal to the cost of one large sewage works. The country badly needs 13 sewage works of this size.

Moreover, according to the PHARE programme rules, spending priorities are defined by the recipient country's government. Thus, the development of the programme in the field of the environment reflects primarily the changing priorities of the Czech government. Incidentally, this also gives supporting evidence for the 'periodization' of environmental concern in the Czech Republic advanced earlier, especially when one allows for the long gap between the submission of project details in Brussels and the receipt of funds.

Initially, projects supported by the PHARE programme in the Czech Republic were aimed at problems related to water and air pollution, waste management, nuclear safety, drinking water and environmental education. The emphasis was later shifted towards 'regional' and 'trans-boundary' projects. The Czech Republic is involved in two such projects: the Elbe project and a scheme about air pollution in the

'Black Triangle', the region on the German, Polish and Czech borders.

The Role of the European Union in Shaping Czech Environmental Policy

It is wrong to expect that the role of the EU in shaping Czech environmental policy would be similar to that in the financial sphere. There has not been anything like a one-way flow of strategies and policies from the EU to the Czech Republic analogous to the flow of money. The periodization of environmental concern in the Czech Republic which, as shown above, was reflected in the government's environmental policy performance, was manifest in the international sphere too. Two periods can be distinguished. Firstly, with a certain degree of exaggeration, we can state that Czechoslovakia was in the forefront of European environmental struggle during the 'enthusiastic' and 'high tide' period of environmental concern. Proposals advanced by the Czechoslovak federal minister for the environment, Josef Vavrousek, and promoted by the government, far exceeded, in terms of their intellectual depth and spatial extent, any strategies developed at that time by the EC. The first pan-European conference of environmental ministers taking place in Dobris near Prague in July 1991 was a direct consequence of an ambitious (and perhaps idealistic) post-revolutionary approach to the environmental problems of the European continent. Almost all European environmental ministers as well as ministers from Canada, Japan, and the US, took part in this conference. The conference goals were:

> . . . to upgrade substantially the existing European environmental protection and restoration institutions, national as well as international, and to integrate them into a pan-European system of co-ordinated 'mechanisms' of environmental efforts at the continental level . . . The second Dobris objective was to develop, implement and then periodically to revise an Environmental Programme for Europe . . . But the third objective was the least conventional and, in my view, the most important. I wanted the Ministers to start to discuss human values and environmental ethics for sustainable development as the basis for such ways of life which can re-establish harmony between humankind and Nature [*Vavrousek, 1993: 92–3*].

These ambitious goals, which have not been lived up to for various reasons, provide evidence of the domestic origin of environmental strategies and policies carried out immediately after the political take-over and developed by environmentalists already under the communist regime. Early 'action programmes', such as the 'Rainbow Programme' and the 'Strategy to care for the environment in Czechoslovakia' (1991), were also based on domestic analyses and experience.

The newly prepared (post-1992) environmental policy is also marked by specifically Czech features, although of a profoundly different kind, and is inspired only to a limited extent by international influences. The opinions of the Czech government on such issues as integrating environmental considerations into other sectoral policies and taking a share in responsibility for the state of the global environment might differ from the official EU view. Even the principle of sustainability is not apparently taken for granted: 'Governmental environmental policy is not taken out of an overall context of social transformation and is not framed within some out-of-touch-with-reality concept of sustainable development, but is aimed at protection and improvement of the environment within the scope of our capacities' (Vaclav Klaus, Czech Prime Minister).[4]

If we turn to environmental policy in a narrower sense, essentially meaning legislation, the situation is importantly different. Harmonization of environmental law and standards remains a task of high priority for the Czech government. The motivation behind this is clear; a main strategic aim of the government is the country's integration into the EU and this requires legislative compliance, including environmental law. Whether this is feasible without acceptance of the underlying principles remains an unanswered question. It should be noted also that the EU seeks Czech compliance as much for the EU's sake as out of a regard for environmental quality in the former Czechoslovakia [*Kara, 1992: 192*].

WHAT HAS BEEN AND WHAT IS TO BE DONE (SUMMARY AND OUTLOOK)

Although many reforms are still unfinished, it seems that there has been more progress than failure (unless one is measuring against the yardstick of initial expectations). What is particularly remarkable is the new legislation as a basic precondition for the reversal of past trends and substantial environmental improvement. After adoption of the new Forest Act and revised Water Act, the legislative framework will be more or less complete. But the harmonization of standards, legislation and practice with those of the EU will be a much longer process, as will be law enforcement – made more problematic by the diminution of the state's role.

A lot of expectations were attached to foreign assistance; it came, but mainly in the form of know-how and advice. Foreign investment in the environment amounted to only 1.5 per cent of the total in 1992. Moreover, the experience has been gained that 'foreign investment is not always environmentally friendly and tends to fix (or even promote) the inherited, undesirable industrial structure characterized by high energy and natural resource consumption per unit of production' [*Stoklasa,*

1992: 21]. Another important area calling for cross-sectoral co-operation is the problem of responsibility for past environmental damage in privatized properties. In fact, the absence of efficient co-operative mechanisms integrating environmental concerns into decision-making is perhaps the main failure of the past three years. Only partial steps in this direction have been made so far and, as a result, environmentally-oriented action is prompted in a rather haphazard way through other mechanisms, for example, as one of the priorities for agriculture grant allocation.

The principles of energy policy – crucial for environmental policy in the Czech Republic – remain somewhat dim and unelaborated. The prevailing approach seems to favour maintenance and even extension of existing capacities (the construction of the nuclear power station Temelin has recently received the go-ahead) rather than the systematic promotion of energy-saving programmes.

CONCLUSION

Our account of the development of Czech environmental policy has highlighted processes working at the domestic and international levels: a close link at the domestic level between degrees of public concern and environmental policy performance, and processes at the international level which have tended to run counter to declining public environmental concern in the Czech Republic.

Going back beyond the autumn 1989 political upheaval, we emphasized the role of environmental degradation as a source of social unrest in the last period of the communist regime and as a 'disguise' which allowed the expression of opposition to the regime. Perhaps unlike the majority of other former communist countries, due to the effort of several personalities, environmental concern in Czechoslovakia went further than just political opposition and was projected into the preparation of strategies and plans for coping with environmental deterioration when the time came. This explains why the reform of environmental policy took place so rapidly after November 1989. Although the 'enthusiastic' period was relatively short, it was enough to enable environmentalists to get a share in power and to realize a proportion of their ideas.

Additionally, 1989 was an international turning point. Before that year differences between West and East resulted in differential treatment of the environment and widely varying political, economic and social mechanisms for environmental policy. Post-1989, Czechoslovak environmental reform laid the ground for overcoming the international policy divide and for convergence with the West. In other words, it prepared the way for harmonization with western and primarily EU standards. There have been two phases of Czech(oslovak) efforts at harmonization, differing in intellectual depth and geographical scope, and this is where the link between the two concepts lies.

The first post-1989 Czechoslovak environmental policy-makers regarded environmental reform as embedded in a general social and political reform, valid for and needed by the whole continent:

> So a major lesson of the 1989 Revolution has been the need to combine appropriate western 'know-how' and technology with a fierce desire for an ecologically sustainable improvement in material well-being and a proper grounding of our intellectual life in honest soils. From this mixture, we may produce a model upon which other Europeans, West and East, may reflect [*Urban, 1990: 135*].

For a certain period, Czechoslovakia carried out an ambitious and progressive environmental 'offensive' internationally. With the arrival of a new group of policy-makers after the 1992 elections, environmental policy has changed both domestically and internationally towards a less ambitious and a more pragmatic form.

Although the main task in the field of international co-operation remains the same – convergence with EU policies – the underlying reasons for pursuing this goal are somewhat different. If during the first 'enthusiastic' period Czechoslovakia's effort to comply with western standards was accompanied by an attempt at taking part in the formulation of continental environmental strategies, the aim of the second period was merely to harmonize environmental law with that of the EU because it is necessary in order to demonstrate the readiness of the Czech Republic to join the EU.

Unlike the previous period, the impetus for further implementation of environmental policy in the near future will be almost exclusively external and 'top-down' (governmental) rather than 'internal' and grass-root, deriving from pressure groups and the public. In any event, the new environmental policy will have to be compliant with the EU, and will in all probability lack radical claims, soberly reflecting the demands of socio-economic transition.

NOTES

1. 'Tramping', a leisure activity inspired by books about the life of American backwoodsmen and cowboys.
2. The poll, carried out by the Institute of Sociology of the Academy of Sciences, involved 2000 respondents all over the Czech Republic; 83 per cent of them considered environmental problems to be the most important, the highest score among 13 items.

3. According to the results of the Institute for Public Opinion Research, environmental problems took eighth position in Oct. 1991 and fifth in Feb. 1992.
4. *Lidove noviny* newspaper, 19 Oct. 1993.

REFERENCES

Commission of European Communities, 1991, *Environment sector strategy paper: PHARE*, unpublished document, Brussels: DG-XI PHARE Co-ordination Unit.
Cuth, J., 1992, 'Asociacna dohoda – zaklad spoluprace CSFR s Europskym spolocenstvom v oblasti zivotneho prostredia', *Zivotne prostredie*, No.5, pp.233–5.
Jehlicka, P. and T. Kostelecky, 1992, 'The Development of the Czechoslovak Green Party since the 1990 Elections', *Environmental Politics*, Vol.1, No.1, pp.72–94.
Jehlicka, P. and T. Kostelecky, (forthcoming), 'The Greens in Czechoslovakia' , in D. Richardson and C. Rootes (eds.), *The Green Challenge. The Development of Green Parties in Europe* (London: Routledge).
Kara, J., 1992, 'Geopolitics and the Environment: The Case of Central Europe', *Environmental Politics*, Vol.1, No.2, pp.186–95.
Lamper, I., J. Machacek and Z. Petracek, 1993, 'Nemame pravo poskozovat Zemi' (Interview with Ivan Dejmal), *Respekt*, No.1, p.9.
Moldan, B. (ed.), 1990, *Zivotni prostredi Ceske Republiky* (Prague: Academia).
Stoklasa, J., 1992, 'Ekologie a nas "vstup" do Evropy (nebo vstup Evropy k nam?)', *Mezinarodni politika*, No.6, pp.21–2.
Sujan, I., 1993, 'Ucinnost makroekonomickej politiky v rokoch 1991 a 1992', *Ekonom*, No.1.
Urban, J. 1990, 'Czechoslovakia: the power and politics of humiliation', in G. Prins (ed.), *Spring in Winter* (Manchester: Manchester University Press).
Vavrousek, J., 1993, 'Institutions for Environmental Security', in G. Prins (ed.), *Threats Without Enemies* (London: Earthscan).
Zeman, J., 1992, *Ekologicky program pro CSFR k 5.6. 1992* (Prague: Ustredni ustav narodohospodarskeho vyzkumu).